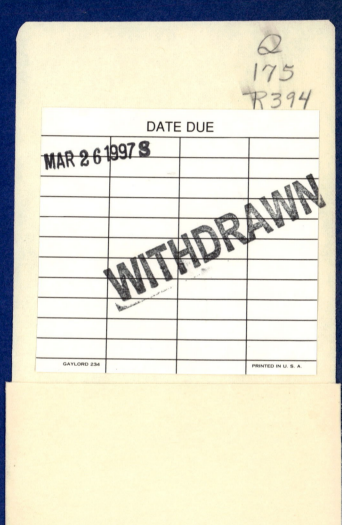

Scientific Explanation

SCIENTIFIC EXPLANATION

by Nicholas Rescher

Research Professor of Philosophy,
UNIVERSITY OF PITTSBURGH

 *The Free Press, New York*

COLLIER-MACMILLAN LIMITED, LONDON

The Free Press
A Division of The Macmillan Company
866 Third Avenue, New York, New York 10022

Collier-Macmillan Canada, Ltd., Toronto, Ontario

Library of Congress Catalog Card Number: 71–80675

printing number
1 2 3 4 5 6 7 8 9 10

The author wishes to dedicate the book to his mother and to the memory of his father.

Preface

This book combines features of three generally distinct genres: the textbook, the monographic treatise, and the polemical tract. It is a textbook in that it seeks to furnish an introduction to basic issues in the theory of scientific explanation. It is a monographic treatise in undertaking a detailed and synoptic survey of various fundamental issues centering around the role of laws in explanation. Finally, it is a polemical tract because it adopts a "point of view" and espouses certain controversial doctrines both as to the nature of explanation itself and especially as to the nature of scientific laws. Throughout, the emphasis of the present treatment of explanation falls upon the role of laws in explanation and, in consequence, on an analysis of the nature of laws.

Some such specialization of emphasis is inevitable in dealing with a topic as complex and many-faceted as scientific explanation. To compensate in at least a small way for such one-sidedness, a reasonably comprehensive bibliography is given to help the reader orient himself in the extensive and ramified literature of the subject. This should considerably reduce the problems that arise for anyone who wishes to pursue the study of topics of special interest to him within this field.

Several persons deserve my thanks for help in connection with the publication of the book. Miss Judy Bazy prepared the typescript, assisted me in compiling the Bibliography, and helped to see the book through the press. Mr. Alasdair Urquhart helped with the bibliography and also with the proofreading. In writing Part III of the book I profited from the critical comments of my colleague Professor Richard M. Gale upon an earlier, draft version. I am also grateful to Professor Peter Brown of the Department of Physics of Duquesne University for information and discussion.

N.R.

Contents

Preface

Analytical Contents

over nature) argued as against the primacy claims advanced on behalf of these various contenders.

7. Explanatory Frameworks and the Limits of Scientific Explanation
   The conception of explanatory framework examined and the variety of such frameworks recognized. The scientific framework is one among many. Its claims to pre-eminence considered. Can science explain everything? Aristotle's regress argument against the explanatory completeness of the scientific framework.

8. The Problem of Explanatory Ultimates
   There is no bedrock category of "ultimate facts" that cannot be explained scientifically. Nevertheless there are certain "ultimate" questions regarding the existence of the world and its nature. These cannot be answered directly in terms of an appeal to "the observable facts" but only indirectly through mediation of the choice of an explanatory framework. The nature of this choice as involving the conscious selection of one among various genuine alternatives, but yet as a choice that is not arbitrary but guided.

APPENDIX I   Are Historical Explanations Different?

Criticism of the claim that the events of human history are unique in a manner that precludes their explanation in the same way in which the sciences explain the occurrences of nature — viz., by subsumption arguments — because the unique is unsubsumable. The types of generalizations operative in history distinguished. The particular relevancy of limited generalizations (restricted in scope, yet lawful) established, and their basis considered. The nonuniqueness of historical events essential to their explanation and interpretation. The problem of the predictability of "historical" transactions considered.

APPENDIX II   On the Epistemology of the Inexact Sciences

The nature of exactness examined. Exactness is not a feature essential to science: there can be inexact sciences. History as an example of an inexact science. The nature of historical laws examined—they are shown to be quasi laws in a special sense. Such laws pose particular problems for the issues of explanation and prediction. The question of a specific methodology of prediction mooted. A consideration of special features of the predictive use of evidence. A rationale is established for the use of experts as predictors. Various aspects of the predictive use of experts are considered and some of the difficulties in this area that pose special tasks for methodological research are outlined.

part I

FUNDAMENTALS OF THE THEORY OF EXPLANATION

1. What Is Explanation?

A speaker may explain *what he meant* by a certain obscure or equivocal statement. An agent may explain *what his intentions were* in performing a certain action. A father may explain to his child *what causes* rain. At the base of such explanations rests a why-question: why he said what he did, why he acted as he did, why it rains, and so on. Throughout, the generic function of such explanations is to rationalize facts: to render them intelligible to a mind seeking to understand. Outside the sphere of *theoretical* explanations of facts there is also a variant category of *practical* explanations, that is, "how-to" explanations that deal with the procedures for performing certain activities (e.g., dancing a waltz), making certain things (e.g., creating a paper airplane), achieving certain objectives (e.g., getting to Singapore) or the like. Such practical, performance-oriented explanations will not concern us at present: our concern here is with the theoretical, understanding-oriented explanations, not with the practical, activity-oriented ones.

It is important to distinguish between *explanation* and *justification*, between giving reasons with respect to some fact at issue

1

why it is true rather than reasons *why we say it is true*. This medieval distinction between a *ratio essendi* ("reason for being") and a *ratio cognoscendi* ("reason for knowing") is crucial. For to answer the question of why some event occurred (say that the earth trembled at noon) I must give an explanation of this fact, but to answer the question of why *I say* that this event occurred, I may merely indicate that this was reliably reported to me by witnesses. In ways such as these one can justify making a statement about something that occurs without giving any explanation of it.

In explaining a fact, we place this fact in the context of others in such a way that they illuminate its existence. The theoretical explanation of a fact is an account of it that deals not merely with aspects of *what* it is, but answers the why-question regarding the circumstance *that* it is, thus rendering its being so intelligible. (It is just here that a *ratio cognoscendi* may prove deficient.) An explanation thus consists of two essential parts or components: (1) the fact to be explained, the *explanandum* as it is often called, and (2) an explanatory account which rationalizes this fact, the *explanans*.

Explanation is sometimes said to consist in "reduction to the familiar." This, of course, relativizes explanation to what *is* familiar. But familiarity is by no means a necessary feature of explanations. One person's familiars are anothers marvels. The Siamese gentleman of the story met with utter disbelief the claim that water on certain occasions became solid. One would certainly suppose that the idea that water solidifies in very cold temperatures could be made acceptable to him, and thus provide the basis for explaining the ice formations in question—but none of this would render the freezing process *familiar* to him. And similarly, when the resident anthropologist explains the native rituals of the tribe to his European visitor, he does not thereby turn this behavior into something familiar to the European. Comprehensibility or *understandability* is the key factor in explanation, not "familiarity."

No hard-and-fast line can be drawn between explanation and description, for there is certainly a category of "purely descriptive explanations" that merely explain *what something is like*: what a toothache feels like, say, or how a charging rhinoceros looks. Such *descriptive* explanations are beside the point of present purposes: our concern here is with explanations that deal with what happens in ways that go beyond mere description.

At another level, however, namely, that of the description of *processes*, description does lie very close to explanation. In a loose sense, of course, one might "understand why" some occurrence took place (e.g., why a certain person got angry when insulted) without understanding the mechanisms involved with *how* this happens. But before we could appropriately claim to have a really adequate explanation of the occurrence one would want to be able to answer in a reasonably full way the *descriptive* question as to the *how* of the occurrence. In this way the relatively complete (or "adequate" or "comprehensive") explanation of an occurrence would certainly call for a great deal of descriptive information (quite apart from any descriptions of the occurrence at issue).

Although explanations in general explicate "facts," actual facts are not alone at issue. While a *real* or *genuine explanation* will give, or purport to give, account of an actual fact, one can also give a *possible* or *potential* explanation of an *assumed* or hypothetical fact. Thus when a child on a moving train asks "What causes the telegraph wires to move up and down?" we "explain" this by remarking that they don't, i.e., by dissolving the fact in question, and explaining not the phenomenon but the appearance.

The question may be posed: "What sorts of things can be explained — what is the potential range of explanatory problems?" The answer is, "Any and all facts whatsoever." The conceivable subjects of explanation therefore exhibit an enormous, indeed a limitless variety. The properties and states of virtually anything, any and all occurrences and events, the behavior and doings of people, indeed every aspect of "what goes on in the world," can be regarded as appropriate objects of explanation. And here "the world" can be taken in the very widest sense, including not only the physical universe, but also the "world of mathematics" and so on.

Explanation is a basic feature of science. Science seeks to organize and systematize our knowledge of what goes on in the world on the basis of explanatory principles that can afford answers to why-questions. In organizing our knowledge of the animal kingdom, for example, we might classify animals in many arbitrary ways — say alphabetically rather than by genus and species. What renders such a classification arbitrary and "unscientific" is just the fact that it has no explanatory value; it cannot be put to work in giving explanations regarding any feature of animate existence, say biological structure or ecological behavior patterns or the like.

2. The Tasks of a Theory of Explanation

A satisfactory theoretical account of explanation must encompass at least four tasks: (1) It should elucidate the linkage or types of linkage that must exist between the explanans and the explanandum if the explanation is to be adequate. (2) It should provide, at least in generic or outline form, the principal considerations bearing upon the correctness and the relative strength of explanations, and so also serve to differentiate between good explanations and poor ones. (3) It should furnish the materials for distinguishing between different types of explanation, providing machinery for the classification of explanations, and bring out conceptually illuminating distinctions between explanations of significantly different sorts; and finally, (4) it should illuminate the place of explanation — and especially of scientific explanation — in the intellectual scheme of things, clarifying its scope and limits, establishing its relationship to cognate procedures such as prediction and retrodiction, and depicting its exact role within the overall project of scientific understanding.

The key entry in this list is clearly the second item: the issue of the evaluation of explanations. When a child contracts measles, this might be "explained" in many ways. For here, as with any given phenomena, there are many different potentially applicable sorts of explanations: the "germ" theory of infection and the occult theory of the evil eye are but two among many ways of accounting for the child's illness. It is obviously a matter of critical importance to settle the question of how the relative correctness of alternatively *possible* explanations is to be assessed. And in particular it is important to examine in a systematic way the exact sources of the superiority of the *scientific mode* of explanation in contrast to other approaches such as the prescientific ("common sense") or the astrological or the occult.

A study of the theory of explanation is of substantial value not only in its own right but because of the light it can cast upon the conceptual structure of science. The task of explaining the phenomena of the world is, after all, one of the main tasks of science — indeed many writers insist upon its being the primary and definitive task. Be this as it may — and the issue is one to be examined in detail below — the clarification of the nature of explanation is bound to bring in its wake a clearer understanding of the nature of the scientific enterprise.

The theory of explanation is a highly interdisciplinary undertaking. It is deeply penetrated by considerations from many branches of inquiry: logic, epistemology or theory of knowledge, and the methodology of science above all. Despite the fundamental and basic character of the concept of explanation — or perhaps because of it — its study is not an elementary discipline but the focal point where many inquiries come together.

3. The Pragmatic Aspect of Explanation and Its
    Abstraction by Science

Some facts strike us as more startling or more significant than others, and it is these above all that we seek to have explained. One does not say "Explain yourself!" to the man who has done what everybody else would have done in like circumstances. In general, explanations have a setting within a *pragmatic context* in which an explanatory question "arises." One does not just explain something, one explains it *to someone* (perhaps only to oneself), so that the explanatory enterprise proceeds within a concrete framework of common inquiry.

When an "explanatory problem" is posed by the consideration of some explanandum, and a suitable explanans is given to provide an appropriate "explanatory answer" to it, the upshot is a completed explanation. The idea of "explaining" is thus basically a dialectical one, involving a question and an answer. At bottom, explanation is a process of communication that envisages a dialogue of interlocutors, one of whom "explains something" *to the other* by rendering this "understandable" to him. Theoretical explanations revolve about questions (why-questions) and this business of question-and-answer is a fundamentally dialectical process.

On this approach, explanation appears as a matter of heuristics — of "rendering something clear to someone" by "putting it into a graspable setting." But in *scientific* explanation this heuristic aspect is attenuated to the point of nonexistence. When we consider the topic of scientific explanation, we largely abstract from this pragmatic aspect of the interlocutor-involving setting within which an explanatory question arises: we imagine (or postulate) an abstract, impersonal framework, rather than a concrete dialectical setting. And we assume a range of questions marked off by abstracted conceptual boundaries of a *discipline* rather than by the personalized range of interests of an individual inquirer. The questions that "arise" are now fixed abstractly (in a subject-matter guided way) by the conventions of the discipline that serves to endow with relative consistency the changeable vagaries of human curiosity. By means of such abstractive conventions science ignores the (undeniably authentic) pragmatic settings in which explanatory questions, including those with which the sciences deal, initially take their root. There is nothing harmful or regrettable about this process of abstraction. By this means, science builds up a highly versatile tool chest of specific explanatory instruments to furnish the instrumentalities concretely needed on various context-specific occasions. Scientific explanation is not a matter of a heuristic designed to make something *understood* by any particular person or group, but of the creation of a context-free body of machinery for rendering things *understandable*. Thus a scientific explanation does not rest upon the background knowledge that an individual may bring to a dialectical situation. It postulates an impersonal, objectified realm of "what is known" (often, no doubt, to precious few). This establishment of an intersubjective realm of facts and generalizations (codified as natural laws) that have become "accepted" is a fundamental feature of the scientific enterprise.

In consequence of this approach, the discussion of scientific explanation casts off any recourse to the social realities of an "explanation" in the sense of an interpersonal exchange between interlocutors. The issue is an abstract one dealing with the logical cogency of certain explanatory arguments, and involves no empirical reference to what facts may be known to certain individuals or what modes of argumentation may strike them as persuasive. The pragmatic, context-dependent aspect of "explanation" in its ordinary sense is put aside, or rather abstracted from by scientific explanation, which takes the character of an essentially logical resource: reasoning to the explanatory conclusion at issue by rigorous principles of inference from premises that provide information postulated as interpersonally known or accepted. Such logical rigor of argumentation is the foundation of scientific explanation.

It might be countered that to say this is to do no more than state a remote ideal at which scientific explanation aims. "After all," it might be objected, "a logical argument is, if valid at all, valid forever, whereas a scientific explanation that justifiably appeared cogent in the days of Lavoisier might well be viewed as extremely deficient by present-day standards." But this objection misses the crucial point. No doubt we would regard many of the plausible explanations of 18th-century chemistry as defective, but we would do so because we reject the explanatory premises they employ, and not because these premises somehow nowadays support the conclusion less effectively than they once did. In this way, scientific explanation calls for arguments to factual conclusions made by timelessly cogent modes of inference from premises vulnerable to the changes of progress reflected in the successive stages in the "state of our knowledge" regarding the workings of nature.

4. What Makes an Explanation Scientific?

Scientific explanations have their starting point in the occurrences of nature — *all* of nature, including man and his works. The first step in science, as indeed in the prescientific, common-sense explanation of what goes on, is the subsumption of observed events under established generalizations. The paradigm of a scientific explanation is given by arguments of a very simple and straightforward sort: "Why did this water boil? Because it was heated to 100°C, and in these altitudes water always boils at that temperature." But science of course does not restrict itself to generalizations of what is encountered familiarly in everyday experience. Its methods of observation are vastly extended; its generalizations are empirically based hypotheses of a complex, highly contrived sort, that have been tested and established (or at any rate, *tentatively* established) in observational situations which may, under laboratory conditions, be extremely artificial.

Science thus strives to explain both the specific concrete occurrences of nature and the general, abstract laws with which it works in explaining such particular occurrences. The basic pattern is essentially the same throughout: occurrences are explained by subsumption under (complexes of) generalizations or laws, and these laws themselves by subsumption under (complexes of) other laws of higher generality. (This last-named tendency builds into science its striving for ever greater generality.)

When giving an explanation in actual practice one may, of course, cite *only* the law that is applicable or *only* the facts that are called for when applying the pertinent law. What explains the door's being open?—Henry opened it. What explains the moon's orbit? Newton's law of gravitation. In the former case no law is explicit, in the latter no specific facts. Such examples, however, do not establish the dispensability of the unmentioned factor: they merely show that something can be left tacit in discussing a context where its presence is more or less obvious. Scientific explanations are invariably subsumption arguments that cite facts to establish an explanandum as a special case within the scope of lawful generalizations.

The explanatory use of *probabilistic* generalizations—rather than strictly *universal* ones—constitutes no real exception to this. If we know that 82 per cent of all male adults exposed to virus X in a certain manner contract a certain disease, then given his exposure to virus X in the relevant manner one can explain Smith's contracting the disease by a subsumption argument with reference to the probabilistic generalization at issue. The nature of the "subsumption" that is at issue in such probabilistic explanations is complex. This will obviously not be a deductive subsumption of the type: Daisies always wilt when parched, and so this parched daisy wilted. But this is a complex and difficult matter to which we shall have to return at considerable length.

The "scientific spirit" calls for the use in explanations of *tested and confirmed* generalizations that have been qualified by the evidence for acceptance as laws. This is a pivotal aspect of science: the use of tests that deploy observational and experimental data, considerations of intersystematic coherence, and the like, as touchstones for the acceptability of hypothetical generalizations. Prescientific explanation, after all, also made extensive use of generalizations. The systematic search for data that might *falsify* — and so also, that failing, to *confirm* — its explanatory mechanisms is a distinctive feature of scientific procedure.

An explanation is thus marked as "scientific" by both (i) its *subject matter* — viz., both what specifically happens in nature (i.e., particular concrete occurrences) and the generic features of natural occurrences (i.e., the laws governing them), and (ii) its *method* — viz., subsumption of the facts to be explained under tested and confirmed laws. The really critical factor is the second one. After all, various occurrences of nature can also be explained by subsumption arguments (no less rigorous in their internal logic than those of the orthodox sort) with reference to principles of a prescientific or nonscientific kind: animism, numerology, astrology, and the like. But the "laws" invoked in such explanations lack that key feature of having survived critically designed tests against experimental or naturally observed data that could and would yield invalidating results.

5. The Role of Laws in Scientific Explanation

To achieve scientific understanding of a fact, we must understand not only *that* it is the case and *what* it consists in, but also *why* it is. Now this is a question we have not succeeded in answering effectively unless our answer puts us into the position of seeing that the other, wrong answers just won't do. Other things being equal, the explanation given for a fact becomes stronger and better according as it succeeds in narrowing the range of alternatives, ruling some of them out as "merely seeming" alternatives, but not genuine or plausible ones.

To explain a fact scientifically is thus to adduce reasons to show why *this* fact obtains rather than some one among its possible alternatives. This requires going beyond establishing that the fact *is* actually the case to showing that (in some sense) it *had to* be the case — that it was *necessary* and inevitable, that it should be so — or at least *probable* and "to be expected." We may refer to this requisite of a scientific explanation as its *modal* aspect. That scientific knowledge demands such a modal condition was already clearly seen by Aristotle, who wrote: "We suppose ourselves to possess unqualified scientific knowledge of a thing, as opposed to knowing it in the accidental way . . . when we think that we know . . . that the fact could not be other than it is."[1]

Now from what source do explanations obtain this *modal* aspect of *necessity* or — at the minimum — *probability*? The answer is that they obtain this from one source alone: their use of laws. For as we have already seen, scientific explanation proceeds by *subsumption under laws*, by placing the item to be explained as a somehow special case within a framework of generalizations that are taken to state how things must operate within a certain range of phenomena. This tree "must" shed its leaves in the fall because it is an elm and all elms, being deciduous, must shed their leaves in the fall. It is this subsumption under laws that endows scientific explanations with their modal aspect, for laws lay claim to a nomic necessity that they can transmit to the particular cases that are subsumed under them.

We can see from this angle of approach how the result of a scientific explanation acquires the aspect of nomic necessity. If the explanatory laws in question purport to state — explicitly or implicitly — how things *must* go in all cases of a certain sort, then, since the scientific explanation takes the form of a subsumption argument, the conclusion of the explanatory argument will also state how matters "must" go. In the application of laws to

1. *Analytica Posteriora* **I**, 2; 71b 8–12.

special cases, a subsumption argument will transmit the necessitarian aspect of the laws to the derivative explanatory conclusion.

The explanatory use of laws is not just a *de facto* actual feature of scientific explanation: it is absolutely indispensable. Only a subsumption-under-laws procedure can achieve the aims that scientific explanation sets for itself: without this recourse to laws, *scientific* explanation would be impossible. The explanatory argument could not endow its conclusion with the requisite modal coloration of necessity (or probability) if it did not invoke universal (or probabilistic) generalizations in an appropriate way. This, of course, merely shifts the problem one step backward for the question remains: Whence do laws get the modal force which they are able to transmit to the special cases subsumed under them in explanation? This question of the source of the nomic necessity of scientific laws is of crucial importance for the theory of scientific explanation. We shall return to this key issue at considerable length in Part III, for the present giving only a brief indication of the direction in which an answer to this question may be sought.

In science, a genuine law — and this is something on which philosophers of science seem to be agreed — must be unrestricted in its applicability by considerations of space, time, and actuality: it must make no reference to specific sectors of actual space-time. "All elms are deciduous" will qualify, but not "All the elms in Smith's yard . . .," or "All the elms existing prior to A.D. 2000 . . .," or "All actually-to-be-observed elms. . . ." A genuinely scientific law will have to eschew any such particularizing references: it will state, or purport to state, not just how things "are" in a certain sector of nature as regards those items specifically at issue, but how they "have to be" in general, in all actual and even hypothetical cases of the relevant sort.

The origin of its modal force lies in the sort of generality that is claimed for a generalization in classing it as a law. In holding that the generalization is *unrestrictedly* applicable to cases of the relevant sort — that it applies without restrictive boundaries confining its potential applicability to limits in space, time, or actuality — we insist it does not only hold for those cases we have examined to date, but to *all possible cases*, specifically including unexamined ones, future ones, and even hypothesized ones. A law is more than a summary of observed experience: it lays claim to a generality of a very strong kind. Yet lawfulness is not to be equated with unrestricted applicability pure and simple. A generalization could be spatio-temporally limited but yet lawful. The critical factor is the crossing of the boundaries of the actual, so that the generalization is endowed with hypothetical force. Even if there are spatio-temporal limitations, the generalization would apply not only to the actual objects (of the appropriate kind) existing within those space-time limits but would be held to apply also to all possible objects within these limits. And just this claim to applicability would render the generalization lawful. (This point will be stressed in the discussion of limited historical laws in Appendix I.) The nomic necessity of laws is a reflection — or, strictly speaking, a consequence — of this strong mode of generality. The "must" or "has to" of nomic necessity is another way of insisting that the law holds good *in all possible cases* of the appropriate sort. To be unwilling to make this claim in behalf of a generalization is to be unwilling to accord it the status of a scientific law.

The development of the argument of the last two sections can thus be summarized as follows. A scientific explanation will be a subsumption argument whose major premiss is a suitably *lawful* proposition. It is a feature essential to lawfulness that, in view of the sort of generality requisite for lawfulness, there be present the element of nomic necessity reflecting how things "have to be" in some sector of nature. A recourse to laws is indispensable for scientific explanation, because this feature of nomic necessity makes it possible for scientific explanations to achieve their task of showing not just *what* is the case, but *why* this is the case. This is achieved by deploying laws to narrow the range of possible alternatives so as to show that the fact to be explained "had to" be as it is, in an appropriate sense of this term.

6. Explanation and the Coherence of Laws

Merely empirical generalizations are inherently unsatisfactory as instrumentalities of scientific understanding. When one only remarks, say, that the rise and fall of a barometer is connected with changes in the weather, that does not go very far for explanatory purposes. Now there is good reason why such an isolated empirical generalization would be relatively inadequate. To provide an *explanation* of something one needs to know not only *that* things are connected in a certain way, but *how* these connections function: one needs some understanding of the *modus operandi* at issue. This is something an empirical generalization by itself is in no position to supply.

Let us consider an example of this. In the history of biology, Darwin's contribution was not the invention of the theory of evolution but its establishment. Prior to his work, evolution was simply a hypothesis, a theory that could be used to account for a body of known biological fact, and that did have some adherents. But when Darwin's work was done, this "hypothesis" had displaced all its competitors and gained acceptance throughout the world of biological studies.

Just exactly what did Darwin do to bring about acceptance of the theory of evolution? Prior to Darwin, there was one central problem which kept evolution on the plane of a purely theoretical possibility, a mere hypothesis: What is the governing mechanism for the evolutionary development of biological species? What could be the guiding force to direct the evolutionary development of a new species from older forms of life? Before admitting *that* evolution occurred, biologists wanted to know *how* it could occur.

To this problem, Darwin gave an answer at once brilliant and decisive. Horticulturists and animal breeders had long realized that it is possible to bring about new biological species possessing desirable qualities through supervised breeding of carefully selected individuals, that is, by artificial selection. Nature, Darwin argued, in fact does just the same, so that we must also recognize the phenomenon of *natural* selection. Darwin shrewdly used Malthus' theory of population to provide the crucial basis for his theory. Malthus had maintained, with impressive evidence to support his view, that human and animal populations will, if unretarded, multiply in geometric proportion, while the supply of plant food on which these populations are ultimately dependent increases in no more than arithmetic proportion. The result, Malthus taught, is a necessary "struggle for survival" among animal populations, resulting from mutual competition for a limited supply of food. Darwin borrowed this Malthusian idea lock, stock, and barrel. It provided the basis for his concept of natural selection by the elimination of the "unfit" and the "survival of the fittest." The animal breeder seeking to reinforce "desirable" qualities selects his strains for guided reproduction. Nature, using the struggle for survival, selects her favorites, the fittest to survive. Drawing on Malthus, Darwin put into place the missing keystone of the governing mechanism of evolution: natural selection. He provided a mechanism for the process of evolutionary development that had heretofore appeared in the light of "merely empirical generalization" as to the development of species.

This example is an instance of a general fact. An empirical generalization is not to be viewed as fully adequate for explanatory purposes until it can lay claim to the status of a law. Now a law is not just a summary statement of observed-regularities-to-date, it claims to deal with a universal regularity purporting to describe how things inevitably are: how the processes at work in the world must invariably work, how things have to happen in nature. Such a claim has to be based upon a stronger foundation than any mere observed-regularity-to-date. The *coherence of laws* in patterns that illuminate the "mechanisms" by which natural processes occur is a critical element — perhaps the most important one — in furnishing this stronger foundation, this "something more" than a generalization of observations. An "observed regularity" does not become a "law of nature" simply by becoming better established through observation in additional cases; what is needed is *integration* into the body of scientific knowledge.

Herein a substantial danger arises for scientific progress, particularly when a new and to date isolated generalization — one that does not fall into the pattern — comes into view. An explanatory mechanism is provided by a theory that is a fabric of interwoven regularities. But if the entire "field" at issue is relatively new, the other threads of the connecting fabric may simply not yet have come to light. Any individual, isolated, empirical generalization will thus be relatively inadequate for explanatory purposes, although that certainly does not mean that there is something intrinsically wrong with it, nor entitle one to doubt its ultimate acceptability. In the 17th century the standard pattern of explanation in physics was conceived in terms of contact interaction. Newton's theory of gravitation, with its apparent espousal of action at a distance, did not fit into this mechanistic scheme. Many scientists of the day were thus distressed about Newton's theory, and, rejecting the idea of a "force of gravitation" as an occult mode of action at a distance, tried vainly to provide a mechanical account of gravitational attraction.

It is, in this regard, worth noting that the bearing of the requirement of coherence is perhaps too easily distorted. The law must certainly fit into *some* pattern, but this need not of course necessarily be *the presently accepted* pattern. It is a convenient but unwarranted step to condemn the unfamiliar as unscientific, and to bring to bear the whole arsenal of scientific derogation (as "occult," "supernatural," "unscientific") that one sees, for example, orthodox psychologists launch against parapsychology.

But the fact that the requirement of coherence for explanatory laws can be abused does not show that it should not be used. (Every useful instrument can be misapplied.) And, of course, the proper use of this requirement must always be conditioned by reference to the primary requirement of correspondence — the evidence of tested conformity to fact.[2]

7. The Grounding of Explanations: Actual versus Possible Explanations

The man who cuts himself when carving the dinner roast might have done so "out of simple carelessness" or "because of anxiety" (being worried about his health) or "because of ill luck" (having walked under a ladder that day).

What is it that differentiates an *actual* or correct explanation from one that is merely *possible* or conceivable? To obtain an answer to this question we must first review the structure of an explanatory argument. Such an argument in explanation of a particular occurrence (say the breaking of a wire) has the following features:

It comprises a set of explanatory premises including (i) certain lawful generalizations (the laws relating the imposed weight to the induced stress upon wires and the resultant tensions created therein, and the laws enabling one to derive from such stresses and tensions the breaking point of the wire), and (ii) certain concrete data to serve as the "boundary-value conditions" for applying these laws (the specific characteristics of the wire, the weight of the stone, the setup of the linkage of wire to stone, and the like). Moreover, these materials must fit together in mutual accommodation to one another: their logical structure must be

2. The topic of "scientific method," dealing with the procedures used in scientific inquiry for the *establishment* of laws, is dealt with only rather briefly in this book, since it lies somewhat abaft of our main topic of the *use* of laws. For an authoritative and lucid introductory discussion, with further references to the literature, see Carl G. Hempel, *Philosophy of Natural Science* (Englewood Cliffs, N.J., 1966).

such that we can derive the explanatory conclusion (explanandum) from them by some plausible mode of reasoning—be it deductive, probabilistic, or whatever.

Now these considerations are, as far as they go, essentially *formal* or structural in nature, they merely insist that the explanandum be linked to the explanatory premises in a certain way without imposing any requirements upon these premises themselves. They leave open the *material* or substantive issue of the *acceptability* or of the *truth* of the generalizations and the boundary-value conditions serving as premises that underwrite the explanatory conclusion. Are these *facts*, or are they only guesses or assumptions? Only when this residual issue is resolved in an appropriate way, and the acceptability of the explanatory premises is established can the potential or possible or candidate explanation at issue come to qualify for acceptance as genuine, as a correct or actual explanation. This sort of "possible explanation" at issue here is just the usual one. When the detective-story protagonist conjectures alternative courses of (hypothetical) developments that could have resulted, given the laws of ballistics, medicine, etc., in the murder, one says that he is thinking of "possible explanations" for the crime in a sense that exactly fits our characterization.

But under what conditions will explanatory premises qualify as acceptable? As regards the data or boundary-value conditions that it uses, it is requisite that they must be actual *facts* or must at the very least be virtually factual. The statements reporting these data must either be known to be true, or must at least be such that with regard to them strong evidence is available to establish their probable truth. As regards the generalizations used in an explanation, they must represent laws that are well established or well confirmed. The essential feature of a true generalization is that it actually be general. Thus for a generalization to qualify as representing a law it must have met successfully the standard sorts of observational and experimental tests and satisfy coherence considerations of the sort discussed. Only generalizations meeting these conditions can qualify under the rubric of "laws" fitted for use in scientific explanations.

An *actual* explanation, therefore, must conform not only to the formal requirement that the explanatory premises, if assumed as hypotheses, will render the explanatory conclusion relatively certain or probable; it must also satisfy the material requirement that the particular premises be fact-asserting (true or highly probable) and that the general premises be law-asserting, and represent generalizations which, being well confirmed, have earned the rubric of lawfulness.

8. The Classification of Explanations

Explanations can be classified and cross-classified according to a variety of quite different principles. The items of the following list are among the most important distinctions that underlie such classifications. We have already considered several of them in some detail.

(A) *Classification according to the type of question posed by the explanandum.* There are (1) explanations of *how to do or to make* something (*practical* explanations), (2) explanations of *what something is like* or also of how it functions (*descriptive* explanations), and (3) explanations of *why* a certain particular event took place or a certain regularity obtains (*theoretical* explanations), etc.

As explained earlier, the third, theoretical category will constitute the focus of our discussion. Explanations of this category will alone be at issue throughout the subsequent classifications.

(B) *Classification according to the kind of objects at issue in the explanandum.* For our purposes the most important of these will be (1) particular events or occurrences — which may then themselves be classified further as natural occurrences in inanimate nature, human actions, etc., or (2) clusters of the items of (1) (e.g., the extinction of a species or the migration of a tribe), or (3) patterns of regularity in the items of (1), such as regularities in nature, in human affairs, in speech, etc.

(C) *Classification according to the conceptual machinery deployed in the explanans (i.e., used in the explanatory account).* (1) One very important class of explanations is the *causal explanation* of events or classes of events. And this category can be further subdivided into sub-categories, there being chemical explanations, mechanical explanations, neurophysiological explanations, etc. (2) A second category is that of the *motivational explanations* of human actions, either in terms of consciously enter-tained reasons, or in terms of motivations that need not necessarily play an overt role in the conscious deliberation of the agents at issue.

(D) *Classification according to the strength of the explanatory link between explanans and explanandum.* The most important distinction here is between (1) *deductive* explanations in which the explanandum follows as a logical consequence from the explanans and (2) *probabilistic* explanations in which the materials adduced in explanans are such as to render the explanandum probable rather than hypothetically certain (i.e., certain relative to premisses).

The domain of scientific explanation can effectively be de-lineated by means of this group of classifications. Typically, such an explanation will have the following features:

1. It poses a *why*-question, a reason-seeking question, rather than a descriptive question of the *what, when,* or *where,* or *how* varieties, let alone a practical one of the *how-to-do* type.

2. It deals as its object with the occurrences in the natural universe, or clusters thereof, or regularities exhibited therein.

3. Apart from the general fact already emphasized that scientific explanation is a procedure of subsumption under laws no more specific *a priori* limitations can be imposed as regards explanatory machinery. However, at any given stage of scientific development certain restrictions may be imposed which will epitomize the understanding people have—at a very general and fundamental level—as to "how things work in the world." Seventeenth-century physicists up to—and indeed beyond—the time of Newton were reluctant to accept action at a distance. Scientists at most periods since antiquity have been unwilling to tolerate explanations in terms of numerological factors, astral influences, and the like.

4. The strength of the explanatory link between the explanatory premisses and conclusion can vary greatly in scientific explanation. Although, of course, preferring satisfactory explanations of maximal—that is to say, of *deductive*—strength, science recognizes that in some contexts these simply cannot be had, and so is prepared to settle, *faute de mieux*, for weaker, probabilistic explanations.

These considerations, then, serve to define and delimit the range of scientific explanations with which we shall be concerned here.

One very basic modern controversy in the theory of explanation relates to the position taken in item 4. Some writers have felt that an explanation cannot possibly be satisfactory if the logical link between explanans and explanandum is not airtight, and so are not prepared to recognize probabilistic explanations as truly scientific. The considerations involved in this issue will presently be canvassed in considerable detail, and a justification will be provided for the position which has been taken rather dogmatically at this juncture.

9. The Idea of "Completeness" in Explanation

Compare the following two answers to the explanatory question, "Why did that liquid turn solid?"

(1) Because that liquid is water, and the temperature fell below 32°F, and all water solidifies when kept at a temperature below 32°F.

(2) Because that liquid was cooled to a very low temperature, and all liquids solidify when kept at a sufficiently low temperature.

What is it that makes the first explanation somehow more adequate and more "complete" than the second? Clearly its greater definiteness — that is to say the greater specificity of its pivotal law, "Water freezes when kept at a temperature below 32°F." As compared with the second explanation this gives more detailed information ("freezing at 32°F") regarding a narrower reference class ("water") in contrast with less detailed information ("solidification at a sufficiently low temperature") for a wider reference class ("liquids"). In going to the second explanation from the first we have traded specificity of context against range of application. This is a small (and perhaps harmless) step towards the more and more encompassing global explanations that verge on the vacuous — that the occurrences at issue happened because "it was the will of God" or because "that is how Fate decreed it."[3] In the scientific explanation of events a loss of precision or definiteness is too dear a price to pay for greater generality or comprehensiveness.

Not that generality is not important — but its place is at another level, that of the explanation of laws. Thus consider the explanation sequence that consists of supplementing (1) by the following explanation of its key premiss: "Why did water solidify when kept at a temperature below 32°F?"

3. Recall the contemptuous reference in Spinoza's *Ethics* (Bk. I, Appendix) to the will of God as "the refuge of ignorance."

(3) Because water is a liquid and for all liquids there is a certain temperature below which they will solidify.

This added explanation provides an importantly illuminating background for the first. In conjunction (1) and (3) have a strength that the vaguer (2), although it is something of a ground-floor counterpart of (3), just cannot match.

There can never be a *complete explanation* of anything for the reverse of the reason that there can never be a complete description of anything. In describing a man we can go on *ad indefinitum* to give greater detail about more and more aspects—the exact shape of his fingernails, lips, and so on. In explaining we can go more and more fully into the reasons why of the reasons why. Thus although one explanation can certainly be more complete than another, no explanation can ever be totally "complete" as such.

Scientific explanations will of course also be incomplete in yet another way, exemplified by the continual progress of science that leads to a constant revision and refinement in our understanding of the laws of nature. The past history of science is such as to support a conjecture of infinite perfectability, of a continual enlargement of the boundaries of understanding and an increasingly larger scale mapping of this region. But the sort of incompleteness that derives from the presumptive *empirical* reality of endless scientific progress is not to be equivalent with the *theoretical* incompleteness we have just been considering. Even if scientific innovation came to a stop because nothing (or nothing of much significance) was ever added to the body of natural law in the domain of scientific knowledge—even if science *in fact* ceased its endless progress and attained a plateau of final "completeness" (farfetched, indeed almost unthinkable, as this might seem)—the *in principle* generality-incompleteness of scientific explanation would yet remain, just as the in principle specificity-incompleteness of scientific description would also remain. In describing we can (in theory) go on indefinitely giving further descriptive details about the descriptive details already given; in explaining we can (in theory) go on indefinitely giving further explanatory details regarding the explanatory details already given. Like description, explanation is an inherently incomplete process.[4]

4. This, of course, does not mean that we cannot reach "the end of the matter" when all our actual questions have been answered. But this "end" is, of course, a practical one dictated by our interests and not a theoretical one. My colleague Alan Ross Anderson tells the delightful story of a boy who asked his mother about something. She told him a little about the matter and concluded by saying that he should ask his father. He replied: "Oh, I don't want to know *that* much about it."

part II

FORMAL ANALYSIS OF EXPLANATORY CONCEPTS

1. The Machinery of Discrete State Systems

To aid in a rigorous exposition of the logical features of explanation, and to make possible a detailed examination of the relationship between explanation and its cognate procedures such as prediction and retrodiction, it will prove very useful to introduce a certain amount of formal machinery. In particular we shall develop a way of describing with mathematical detail, and so with substantial precision, an interesting family of physical systems of a particularly simple kind. This mathematical machinery, albeit very simple, will facilitate a far more detailed and exact analysis of the characteristics of explanation than would otherwise be possible.

We shall now concern ourselves exclusively with physical systems that, at any given moment of time, exhibit some definite, specifiable state of affairs, a state that is not instantaneous but lasts for some interval of time (however short). We thus assume a temporal parameter which is not continuous but *discrete*, so that one has to do with a discontinuous time variable that represents discrete periods (e.g., microseconds, minutes, years). A system of this type, which exhibits some particular state for each of the — perhaps very short — time periods (intervals) at issue, will be termed a *discrete state system*, a DS-system for short. In the event that the system is such that it can exhibit only some finite, limited number of diverse (qualitatively distinct) states, one may speak of a *finite* discrete state system.

The actual existence of DS-systems can be illustrated by numerous examples. One such is an electronic digital computer (or even an office desk calculating machine) that exhibits, during any sufficiently small time period, some particular complex of affairs in the arrangement of its calculating and information-storage components. Another example is an atom initially of some radioactively unstable element, such as Uranium I, which for any very small time period must assume the "state" of being an element somewhere in the Uranium-Lead region of the Uranium-Radium series. Furthermore, even a continuously changing system can be redescribed in DS-system terms. Consider for instance a particle moving in Brownian motion over the surface of a liquid within some container. If this surface is subdivided (conceptually) into a number of subdivisions — labeled 1 through n — and a suitably small time interval selected, we could specify "the state" of this system at time (interval) t as corresponding to the location of the particle in that division in which it spends the most time during the interval in question. By such discretizing one can describe — or redescribe — the actually continuous Brownian motion of the particle as a DS-system, and in fact a *finite* DS-system.

Throughout this discussion, it will be supposed that the DS-systems under consideration are governed by "laws of transition" with respect to their states — laws specifying that whenever the "present" state of the system is such and such, then the "next" state (or group of possible "next" states) will be so and so. Such laws of transition governing the behavior of a DS-system may be either *deterministic* or *indeterministic* (probabilistic). A deterministic law is one of the form, "State X is always and invariably followed by state Y." An indeterministic (probabilistic) law will have a form such as, "State X is followed by state Y with probability p, and by state Z with probability $1-p$." Throughout the subsequent discussion, we shall assume that the probabilistic laws under discussion satisfy the "Markov property,"[1] in the sense that the probability that the system will assume state X at time t is a function only of the state of the system at time $t-1$, and is wholly independent of the history of the system prior to time $t-1$. The behavioral theory of systems conforming to these general requirements is studied in probability theory and mathematical physics under the chapter heading of *Markov chains*.[2] A great variety of physical processes have been represented and studied from this point of view.[3]

It will prove expedient to restate these considerations in a more rigorous formalized way, and to introduce some further machinery for the discussion of discrete state systems.

The physical systems now at issue can exhibit at most some countable number of states: $S_1, S_2, S_3, \ldots$ It is assumed further that time is discretized into intervals of fixed size. Consequently, one can portray the history of the system under consideration by the finite list of its successive states throughout the time span at issue, say from mth to the nth interval, giving this history by the list

$$st(m), st(m+1), \ldots, st(n-1), st(n)$$

where $st(i)$ is that member of the set of all possible states $\{S_1, S_2, S_3, \ldots\}$ which represents the actual state of the system for the interval $t = i$.

1. See W. Feller, *An Introduction to Probability Theory and Its Applications*, vol. 1 (New York, 1950), chapters 15 and 16; or J. G. Kemeny and J. L. Snell, *Finite Markov Chains* (New York, 1960), chapter 2.
2. See W. Feller's book cited in the preceding footnote for further reference to the literature.
3. For two especially important examples see P. and T. Ehrenfest, "Über zwei bekannte Einwände gegen das Bolzmannsche H-Theorem." *Physikalische Zeitschrift*, vol. 8 (1907), pp. 311–314; and Ming Chen Wang and G. E. Uhlenbeck, "On the Theory of Brownian Motion, II," *Reviews of Modern Physics*, vol. 17 (1945), pp. 323–342.

Such discrete state systems can be subject to two kinds of laws governing the transition of states, among others. First, there will be *deterministic* laws of state-determination of the type:

$$\text{If } st(t) = \mathbf{S}_1, \quad \text{then} \quad st(t+1) = \mathbf{S}_j$$

Secondly, there will also be *indeterministic* or *probabilistic* laws of state-determination of the type:

If $st(t) = \mathbf{S}_i$, then $st(t+1)$ will be one of the states
$\mathbf{S}_{j_1}, \mathbf{S}_{j_2}, \ldots, \mathbf{S}_{j_n}$, with probabilities $p_1, p_2, \ldots, p_n$ respectively.
(These p_i must obviously sum to 1.)

All the probabilistic laws entering into our discussion are assumed to be of this form, and hence will satisfy the "Markov property" that the conditional probability that state $\mathbf{S}_i$ be succeeded by state $\mathbf{S}_j$ is defined independently of the past history of the system prior to its (assumed) attainment of state $\mathbf{S}_i$.

Any finite DS-system governed by deterministic and/or indeterministic laws of this sort can thus be represented by a square matrix based on the "transition probabilities" of the system. This matrix takes the form

$$\| a_{ij} \|$$

where a_{ij} is the (conditional) probability that if the system is in the state $\mathbf{S}_i$ at time t, then it will be in state $\mathbf{S}_j$ at time $t+1$; i.e., a_{ij} is the probability that $st(t+1) = \mathbf{S}_j$ given that $st(t) = \mathbf{S}_i$. (It is clear that the rows of such a matrix must always add to 1.) This matrix of the transition probabilities for a DS-system will be called the *characteristic matrix* for the system.

A DS-system may be characterized as (strictly) *deterministic* if all of the nonzero elements of its characteristic matrix are 1, and otherwise — i.e., if some nonzero element differs from 1 — the DS-system will be characterized as (at least partially) *indeterministic*.

It will often prove more convenient and more perspicuous to recast the substance of the characteristic matrix of a DS-system in diagrammatic form. Thus consider, for the sake of illustration, the strictly deterministic DS-system with characteristic matrix:

| | | Successor State | | |
		S_1	S_2	S_3
Predecessor State	S_1	0	1	0
	S_2	0	0	1
	S_3	1	0	0

This system is more simply described by the "transition diagram":

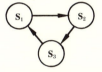

This diagram presents the same information given by the characteristic matrix, but does so in a more readily apprehensible way. Again, consider the indeterministic DS-system specified by the characteristic matrix:

| | | Successor State | | |
		S_1	S_2	S_3
Predecessor State	S_1	.5	.5	0
	S_2	0	.5	.5
	S_3	.5	.5	0

This system is more perspicuously described by the (probabilistic) transition diagram:

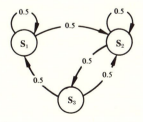

In many cases, a diagram of this sort will specify a DS-system in a manner more graphic, and yet no less exact, than its tabular characteristic matrix.

It is important to recognize the fact that, in considering any finite DS-system on the basis of this abstract mode of characterization, we are not dealing with a merely conceptual possibility. Any finite DS-system can readily be realized as a physical fact through "simulation" on a digital electronic computer of the sort now in existence.[4] Such systems thus represent physical systems that are "possible" not merely in the remote sense of being capable of actual physical realization through the use of instrumentalities actually in existence at the present day.

With this descriptive machinery of "characteristic matrices" and "transition diagrams" at our disposal for the convenient presentation of both deterministic and indeterministic DS-systems, we can now return to the main object of our discussion: the theory of explanation and of such intimately cognate concepts as prediction and retrodiction.

2. Explanation, Prediction, Retrodiction

Before taking up the question of how the concept of explanation and its conceptual cognates of prediction and retrodiction will function in the setting of the special type of physical system with which we are presently concerned, namely the DS-systems, it is advisable to consider the meaning of these concepts as they relate to physical systems in general, quite apart from our specialized standpoint. This essential preliminary discussion is complicated by the necessity for drawing several basic distinctions, especially: (1) that between deductive and probabilistic

4. An unimportant qualification to this sweeping statement is the consideration that a "very large" finite DS-system (i.e., one involving an enormous number of states) may require such an extensive description as to outstrip the memory-storage capacity of existing machines.

explanations and (2) that between actually and potentially explanatory arguments. We must also recognize that, and how, the concepts of prediction and retrodiction are subject to distinctions similar to those of (1) and (2) above. Throughout, we shall be concerned only with the explanation, prediction, and retrodiction of *particular* facts (as opposed to "logical constructs" from individual facts, such as the disjunct "Fact 1 *or* Fact 2" or the like). By a "fact" in this context one is to understand that the system in question exhibits a particular state at a particular time.

1. *Explanation.* A *potential explanation* of the (assumed) fact that a system exhibits the characteristic **Q** takes the form of an argument whose conclusion is that the system exhibits the characteristic **Q**. The premises of this argument consist of two types of statements; namely (1) the general laws, $L_1, L_2, \ldots, L_m$, governing (i.e., known or assumed to govern) the behavior of the system in question, and (2) certain boundary statements to the effect that the system exhibits the characteristics $C_1, C_2, \ldots, C_n$, all different from **Q**.

It should be stressed that by a "characteristic" of a physical system we do not mean simply a (timeless) *property*, such as "having a weight of 15 grams," or "being 24 cm. in height"; for if this were *all* that is meant, a system could both exhibit such a property (at one time) and also fail to exhibit it, or exhibit some contrary property (at another time). Thus a "characteristic" must here be understood as involving not just properties or relations taken in an abstract and timeless way, but also taken to involve the specific times at which these properties qualify the system. This is essential if the law of contradiction is to apply to the "characteristics" of physical systems, as it must for the purposes of our discussion. Consequently, a phrase such as "has the characteristic **C** at time t," although redundant, has the convenient merit of rendering explicit the temporal reference of the "characteristics" treated in our discussion.

A potentially explanatory argument becomes an (*actual* or *correct*) *explanation* if its factual premises are really true and its general premises genuinely lawful.

2. *Prediction.* A *potential prediction* of the supposed fact that a system *will* exhibit the characteristic **Q** at time *t* is an argument whose conclusion is the statement that the system exhibits **Q** at *t*, and whose premises consist of two types of statements: (1) general laws, $L_1, L_2, \ldots, L_m$, governing (i.e., known or assumed to govern) the behavior of the system, and (2) data statements to the effect that the system exhibits the characteristics C_1 at t_1, C_2 at $t_2, \ldots, C_n$ at t_n, *where all of the times involved are anterior to t*, i.e., all $t_i < t$.

A potentially predictive argument becomes an (*actual*) *prediction* if its factual premises are really true and its general premises genuinely lawful, and moreover (i) all the times at issue in the "data statements" antedate the present, so that its type-(2) premises — see item (2) of the formulation given above — give (true) information, but solely about the relative past characteristics of the system (i.e., all $t_i < t_N$, where t_N is the time "now," that is to say, *the time at which the explanation is proffered*), and (ii) its conclusion relates to the *relative* future (i.e., $t_N < t$, where t_N is the time "now").

This analysis of the concept of "prediction" does not accord with the mode of construction placed on this concept in recent discussions by Professor Carl G. Hempel, with the inevitable result that major discrepancies arise in what is said about prediction under these divergent interpretations of the concept.[5]

5. See C. G. Hempel and Paul Oppenheim, "Studies in the Logic of Explanation," *Philosophy of Science*, vol. 15 (1948), pp. 135–175, and especially Section 6 of Hempel's "Deductive Nomological vs. Statistical Explanation" in H. Feigl and G. Maxwell (eds.), *Minnesota Studies in the Philosophy of Science*, vol. 3 (Minneapolis, 1962). Also, Section 3 of Adolf Grünbaum, "Temporally Asymmetric Principles, Parity between Explanation and Prediction, and Mechanism versus Teleology," *Philosophy of Science*, vol. 29 (1962), pp. 146–170, should be consulted regarding the terminological distinction here at issue.

I find it necessary to dwell briefly on the justification of this perhaps "merely" semantic but nevertheless important point of difference.

In accordance with our discussion above, a *prediction* of the supposed fact that a system will exhibit the characteristic **Q** at time t is an argument which bases this conclusion upon premises of the following two types: (1) laws governing the behavior of the system, and (2) true (or assumedly true) data statements to the effect that the system exhibits the characteristic C_1 at time t_1, C_2 at time $t_2, \ldots$, and C_n at time t_N, where

(a) $t_i < t$, for all i (i.e., all premises deal with times *earlier than the time at issue in the conclusion*).

(b) $t_i \leq t_N$ for all i (i.e., all premises deal with times *no later than the present*).

(c) $t_N < t$ (i.e., the conclusion relates to the future).

Now Professor Hempel's construction of "prediction" differs in effect from the foregoing characterization in that he drops conditions (a) and (b). I shall consequently attempt to justify these conditions, and to explain why I find it necessary to include them as a part of the analysis of what constitutes a scientific prediction.

Let me begin by observing that condition (a) is an immediate logical consequence of conditions (b) and (c) taken jointly. It thus suffices my purposes to justify condition (b) as an essential part of an appropriate construction of "prediction."

Assume that condition (b) were violated, so that some premiss asserts that the system exhibits characteristic C_j at time $t_j > t_N$, i.e., for some *future* time. Now how could we possibly go about justifying the use of this futuristic premiss in a predictive argument? There are but two possibilities:

(I) The statement that the system will exhibit C_j at t_j can be derived (be it deductively or probabilistically) on the basis of information about the present and past characteristics of the system, or

(II) The statement that the system will exhibit C_j at t_j cannot be derived in this way.

But note now that in the first case, one effectively carries the predictive process back to a (reformulated) prediction that conforms fully to our requirements, specifically including condition (b). However, in the second case, we do not actually have a proper prediction at all—for the "predictive" argument in view is based on a premiss which cannot be justified in terms of *available* information. In this case one does not have a proper "prediction" at all, but (at best) a *conditional* or *hypothetical* prediction that makes essential use of an undefended and indeed currently indefensible assumption regarding the future characteristics of the system under discussion.

The foregoing line of reasoning may be summarized in the dilemma that an Hempelian prediction *either* does not constitute an argument deserving to be labeled as a scientific "prediction," *or* it can be formulated as a prediction conforming to the requirement laid down in our foregoing discussion.

In summary then, the conception of "prediction" that results from deleting the conditions (a) and (b) of our analysis is, on our view, faulty as a construction of prediction proper. It overlooks the factor that a predictive argument is *not* simply one whose conclusion makes some claim about the future (based possibly on premisses regarding the still more distant future). Rather, in predictive reasoning, as we view it, there should be a temporal structure in the chronological direction of the argument itself, in that it moves from temporally anterior premisses to a temporally posterior conclusion.

3. *Retrodiction.* A *potential retrodiction* of the purported fact that a system has exhibited the characteristic **Q** at time *t* is an argument whose conclusion is the statement that the system

exhibits **Q** at t. The premises of this argument consist of two types of statements; namely, (1) the general laws, $L_1, L_2, \ldots, L_m$, governing (i.e., known or assumed to govern) the behavior of the system, and (2) statements to the effect that the system exhibits the characteristics C_1 at t_1, C_2 at $t_2, \ldots, C_n$ at t_n, where all of the times involved are posterior to t, i.e., all $t_i > t$. Again, with retrodiction, as with prediction, the temporal pattern of the premissed information is a key factor.

A potentially retrodictive argument becomes an (*actual*) *retrodiction* if (i) its premises are actually true (or assumedly true), and (ii) its conclusion relates to the *past* (i.e., $t < t_N$, where t_N is the time "now").

It follows from the foregoing specifications that whenever a prediction or retrodiction (of whatever type, actual or potential) is given, so *a fortiori* is an explanation (of the corresponding type). Our defining conditions for prediction and retrodiction have in effect added to the conditions for explanation (in which the element of time does not function) certain further restrictions of a temporal character. In this light, both predictions and retrodictions appear as explanations of a certain sort.

* * *

The preceding analysis of explanation, prediction, and retrodiction has addressed itself to considerations regarding physical systems in general. We now again focus our discussion upon our DS-systems. From this narrowed perspective we can see that the foregoing general concepts of explanation, prediction, and retrodiction bear upon DS-systems in a special way. We shall henceforth speak only of *potential* explanation, prediction, etc., since the manner of differentiating between potential and actual explanation, etc., is not affected.

The general concepts of explanation, etc., as previously discussed, bear upon the narrowed context of DS-systems in a somewhat modified way, for reasons now to be explained. It is usual in discussions of the methodology of science to consider the "laws" that constitute the machinery for explanation, etc., under two heads:

(i) *Laws of succession* which relate characteristics of a physical system at some particular time with its characteristics at *earlier* or *later* times (e.g., Kepler's laws of planetary motion, or Galileo's law of falling bodies).

(ii) *Laws of coexistence* which relate different concurrent characteristics of a physical system with one another (e.g., Hook's law relating the mass supported by a spring with its elongation; or Ohm's law relating the electrical resistance, potential difference, and current of an electric circuit; or Boyle's, Charles' and van der Waals' laws for gases which relate the pressure, volume, and temperature parameters of a gas).[6]

Now this general distinction regarding laws governing physical systems does *not* carry over to our DS-systems. For *all* contemporaneous characteristics of a DS-system (i.e., all that are being considered when the system is described in DS terms) are fused together into one all-embracing characteristic — namely, "the state" at the given time. Thus only laws of succession are applicable here, laws of coexistence play no role in the discussion of DS-systems. This has the consequence that certain issues that arise in the general theory of scientific explanation do not appear within the more limited domain of the present discussion. And in particular the concepts of explanation, prediction, and retrodiction bear upon DS-systems in a somewhat simplified way, shortly to be set forth.

It is thus clear that, due to the specialized character of DS-systems, one cannot argue validly from a finding that some result is necessary (i.e., always to be found) in the special case

6. I take this distinction between laws of succession and coexistence from Section 4 of Hempel, *ibid.* Hempel treats this distinction as well known; it functions in Mill's *System of Logic*, but goes back at least to Comte. See K. R. Popper, *The Poverty of Historicism* (Boston, 1957), p. 116. (I owe this reference to Professor Hempel.)

of DS-systems to the conclusion that this result is necessary in general (i.e., is always to be found, even apart from our special limitation). But on the other hand, we can indeed argue validly from the finding that some result is possible in the special case (i.e., that some circumstance is realized with respect to certain DS-systems) to the conclusion that this state of affairs is unqualifiedly possible. This mode of reasoning represents a strategy of argumentation that will be used extensively in the ensuing discussion. By exhibiting the sorts of situations that can arise with respect to explanation and its cognates in the special case of DS-systems, we shall explore basic facets of the workings of these concepts.

3. Deductive versus Probabilistic Explanatory Arguments

All the three types of reasonings considered in the last section have two distinct forms, deductive and probabilistic. The distinctions at issue here must be developed in detail.

1. *Explanation with DS-Systems.* In general, a potentially explanatory argument can fall into one of two categories. It may be either *deductive* (*D*-explanation), or *probabilistic* (*P*-explanation). With a deductive explanation, the explanatory premisses would, if true, provide *conclusive* evidence for the conclusion, constituting a *totally sufficient* guarantee of the explanatory conclusion. With a probabilistic explanation, the explanatory premisses do not provide a guarantee of the conclusion, but merely render it relatively likely, and so endow it with a relatively substantial (conditional) probability, say, one in excess of one half, or perhaps some other specified value k in the interval $0.5 < k < 1$. In the special case of DS-systems, an *explanatory argument* is one that concludes that $st(t) = \mathbf{S}_i$ on the basis of an identification of a group of states $st(t_1)$, $st(t_2)$, . . . , $st(t_n)$, where all of the $t_i \neq t$. The argument may be either deductive (D-explanation), or probabilistic, either in the strong

sense (P_s-explanation) that $st(t) = S_i$ is (conditionally) *more likely than not*, or in the weak sense (P_w-explanation) that $st(t) = S_i$ (conditionally) *more likely than $st(t) = S_j$* for any particular $j \neq i$ (i.e., is relatively more likely than its competitors taken singly, though not necessarily more likely than not, or, equivalently, more likely than its competitors taken collectively).[7]

2. *Prediction with DS-Systems.* In general, a potentially predictive argument can also fall into the deductive or the probabilistic type. With a *deductive* prediction (D-prediction), the predictive premisses must, if assumed to be true, provide *conclusive* evidence for the predicted conclusion. With a probabilistic prediction, the predictive premisses do not provide a conclusive guarantee of the conclusion, but merely render it relatively likely, and so endow it with a relatively substantial probability. In the special case of DS-systems a *predictive argument* is one that concludes that $st(t) = S_i$ on the basis of an identification of a group of states, $st(t_1)$, $st(t_2)$, ..., $st(t_n)$, where all the $t_i > t$. The argument may be either deductive (D-prediction), or probabilistic, either in the strong sense (P_s-prediction) or in the weak sense (P_w-prediction).

3. *Retrodiction with DS-Systems.* In general, a potentially retrodictive argument may be either deductive or probabilistic along lines exactly parallel with those envisaged in the case of explanations and predictions. In the special case of DS-systems, a *retrodictive argument* is one that concludes that $st(t) = S_i$ on the basis of an identification of a group of states, $st(t_1)$, $st(t_2)$, ..., $st(t_n)$, where all $t_i > t$. As with prediction above, we again

---

7. This weakest sense of "explanation" in the context of DS-systems does not fully conform to the general concept of "explanation" as presented under heading 1, and it may indeed be too weak to answer fittingly to what we mean by an "explanation" in ordinary, informal discourse. But it is hardly worth making much of this purely terminological point; and at any rate what we mean by a P_w-explanation is both clear in itself and reasonably analogous to our general conception of "explanatory" reasoning.

have three possibilities: D-retrodiction, P_s-retrodiction, and P_w-retrodiction.

It deserves notice that a retrodictive argument may involve predictive subarguments in such a way as to render the entire process of retrodictive reasoning quite intricate. Consider, for example, the DS-system[8]

	S_1	S_2	S_3	S_4
S_1	0	.1	.9	0
S_2	0	0	0	1
S_3	0	0	0	1
S_4	1	0	0	0

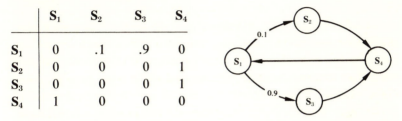

Suppose it is given that the "present" state is S_4, and suppose a retrodiction of the preceding state to be asked for. Superficially it seems as though from S_4 we simply could not judge between S_2 and S_3 as the superior candidate for its immediate predecessor. But notice that we can D-retrodict that S_1 had to be a 2-interval predecessor of S_4. And from S_1 we can P_s-predict S_3 as the "next" state — i.e., we are far more likely to get from S_1 to S_4 via S_3 than S_2. Thus we are, after all, able to P_s-retrodict that S_3 is the immediate predecessor of S_4. The presence of a deviant subargument has no effect on the classification of the argument as a whole, which is determined by the temporal structure of the movement from premises to conclusion, not by the deviations of the path through which this movement is accomplished.

Whenever a prediction or retrodiction (of whatever type) is given, so *a fortiori* is an explanation of the corresponding type; though the converse will not, of course, be true in general. Another, quite obvious, result is that whenever an argument satisfies the conditions for one of the stronger types of explanation (etc.) the conditions for the weaker counterparts are met *ipso facto*.

8. I owe this example to F. Brian Skyrms.

Since our major concern will be with the three classes of DS-specialized conceptions we shall henceforth simply speak of them as *"explanation,"* *"prediction,"* and *"retrodiction"* when these terms are used unqualifiedly. Thus, in particular, we shall be construing these terms in their *potential* sense.

These brief considerations as to the deductive and probabilistic varieties of explanation and the cognate procedures of prediction, and retrodiction — both in general and as specifically applicable to DS-systems — should suffice our needs for the purposes of examining some of the ways in which these concepts can function in their workings in the context of physical systems. We turn now to such an examination.

4. Explanation and Its Cognates in Deterministic DS-Systems

The following thesis embodies what unquestionably is, from the standpoint of the theory of explanation and prediction, the main feature of strictly deterministic DS-systems:

(T4.1) *In a strictly deterministic DS-system, D-prediction, and a fortiori D-explanation (and thus the weaker modes of explanation and prediction), are always possible. In other words, given the state of the system at time t one can always — by the laws of the system — deduce its state at time t + 1, and can thus always D-explain the state at any time t by reference to that obtaining at time t − 1.*

The proof of this result proceeds along obvious lines. The significance of this thesis, which indicates the definitive characteristic of deterministic systems, is too obvious to warrant discussion. A second important characteristic of DS-systems is given by the thesis

(T4.2) *In a strictly deterministic DS-system, D-retrodiction may be impossible: the laws of the system may fail to render*

it possible to deduce from the state of the system at time t what was its state at a prior time.

DISCUSSION.   The thesis is most readily established by a concrete example. Consider the (strictly deterministic) DS-system

	S_1	S_2	S_3
S_1	0	1	0
S_2	0	0	1
S_3	0	1	0

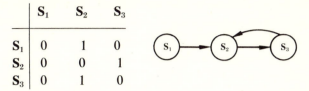

Suppose it is given that the "present" state of the system is $st(t) = S_2$ (and that thus its entire subsequent history is implicitly given). It is nevertheless clear that we are unable to deduce the state of the system at time $t-1$, for we cannot rule out either of the possibilities $st(t-1) = S_1$ or $st(t-1) = S_2$. (Note, however, that the latter is the more likely since the system can exhibit S_2 with unlimited repetition, but can exhibit S_1 only once; so that P_s-retrodiction is indeed possible.)

It should be noted that *any* finite strictly deterministic DS-system that is not wholly cyclic exhibits the essential feature of the foregoing example, namely, a "return" from the "terminal" state of the transition diagram to some prior state, which thus inaugurates a subcycle from whose "initial" state (having two arriving arrows) there *cannot* be any D-retrodiction. This feature has the consequence that all-out D-retrodiction is necessarily impossible in all finite DS-systems, deterministic as well as indeterministic, except for the purely cyclic.

On the other hand, P-retrodiction will inevitably have to be possible in a strictly deterministic *finite* DS-system, because its state-succession history must have the structure of a beginning sequence followed by a cyclic repetition. This means that the cycle-inaugurating state with respect to which the impossibility of D-retrodiction arises (S_2 in the example) is far more likely to result out of the "terminal" state of the infinitely repetitive cycle (S_3 in the example), than out of its unique predecessor in the beginning sequence (S_1 in the example).[9]

The question must also be considered whether the weaker probabilistic modes of retrodiction are always possible in strictly deterministic infinite DS-systems. The negative answer to this question is established by the example of the following system.

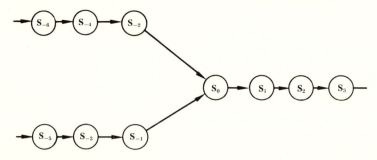

When the history of the system reaches S_0 (as it must) even the probabilistic retrodiction of any of its predecessor states is impossible.

A whole host of important considerations revolve around the issue of the temporal structure of explanatory reasoning (and its cognates). When speaking of temporal considerations it is clear that it is not the laws which are at issue — for they are constant and as it were timeless — but the data relating to the transient states of the system.

Certain systems will be such that given a state of the system (*any* state) — say that at t_0 — one can explain this state by means of the known laws of the system in terms of some (finite) body of data about the states of the system prior to t_0. Such a state will be said to be "*a parte ante* explicable" in the apposite mode of explanation. An occurrence of a state in the history of the system can be explained in this manner if and only if this occurrence was *predictable*. A system all of whose states throughout its

9. This discussion requires a shift in our standpoint regarding probabilities — for they are now not a matter of the *intra-systematic* internal processes of functioning of the system, but of the *extra-systematic* behavior of the system as revealed in external observation.

entire history are *a parte ante* explicable can be characterized as such *as a whole.*

Note that we shall have the thesis

(T4.3) *A DS-system all of whose possible state occurrences are* a parte ante *D-explicable must be strictly deterministic.*

DISCUSSION. Suppose that two (or more) arrows ever departed from one state, as in the case

Then the occurrence of S_j would then not be D-explicable *a parte ante.* Thus no state can have more than one departing arrow, and the system is in a consequence strictly deterministic.

Moreover, we shall even have the cognate thesis

(T4.4) *A DS-system all of whose possible state occurrences are* a parte ante *p-explicable must also be strictly deterministic.*

DISCUSSION. Suppose again that two (or more) arrows ever departed from one single state, as in the case considered just above. Then one of the arrows must depart from a probability entry less than and equal to another, so say $a \leqslant b$. But then suppose that on some occasion S_i is in fact succeeded by S_j. Then this occurrence of S_j is something that cannot be P-explained *a parte ante.*

A system in which some possible state occurrences are not *a parte ante* explicable might be said to possess "an open future." At least some of its states cannot possibly be predicted and be explicable at best *a parte post.* It follows from (T4.2) that every system that is not strictly deterministic will be of this sort. In such a system some occurrences can be explained only *ex post facto* with "the wisdom of hindsight": they could not, even in principle, have been predicted in advance.

The temporal structure of data utilization in predictive reasoning poses no novelties, for the simple reason that, as was brought out in previous discussion, such temporal inference has to be considered an integral part of the very concept of prediction. However, the general importance of finiteness and discreteness is worth noting in this context. We do not want to say that the state of the system at t_0 can be predicted, if it can be determined only if one knows the state at *every* preceding interval, or the state at $t_0 - 1$, and at $t_0 - \frac{1}{2}$, and at $t_0 - \frac{1}{4}$, and so on. Only if laws plus a *finite* body of (temporally prior) data suffice for the state determination is the situation a genuinely predictive one.

5.   Explanation and Its Cognates in Indeterministic DS-Systems

We turn now to a consideration of the status of explanation, prediction, and retrodiction in the context of indeterministic DS-systems. The relevant epistemological issues come to be operative here in a significantly more complex, and thus perhaps all the more interesting, way.

Doubtless the most striking feature of indeterministic DS-systems — in view of the fact that the behavior of such systems is, after all, fully and entirely specified in terms of given laws — is contained in our first thesis.

(T5.1)   *In an indeterministic DS-system, P_w-explanation — and a fortiori both stronger types of explanation, as well as all types of prediction and retrodiction — may be uniformly (i.e., for all states) impossible.*

DISCUSSION.   The thesis is most simply established by a concrete example. Consider the system

	S_1	S_2
S_1	.5	.5
S_2	.5	.5

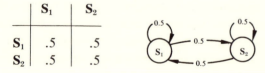

Notice that even if the complete history of the system except for one "unknown" state is given, say

$$\cdots S_2S_1S_1S_2 - S_1S_2S_2S_1 \cdots$$

one cannot (not even probabilistically) fill in the blank by S_1 rather than S_2, or conversely. Thus if we are given that the blank position is actually S_2, we are unable to give even a P_w-explanation of this fact.

Much interest also attaches to the cognate thesis

(T5.2)   *In an indeterministic DS-system, P_w-explanation — and a fortiori both stronger types of explanation, as well as all the types of prediction and retrodiction — may be selectively (i.e., for some states) impossible.*

DISCUSSION.   Again, an example serves to prove our point. Consider the system

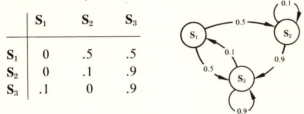

	S_1	S_2	S_3
S_1	0	.5	.5
S_2	0	.1	.9
S_3	.1	0	.9

Even if the complete history of the system is given, except for one unknown state, say

$$\cdots S_3S_1 - S_3S_1 \cdots$$

one may (as in this case) not be able to fill in this blank, not even probabilistically. Should it be the case that the blank is S_2, we are unable to give even a P_w-explanation of this fact. Nor can an occurrence of S_2 ever be (even weakly) predicted or retrodicted. The occurrence of this state lies not outside the realm of possibility but, so to speak, outside the range of scientific rationalization.

It is perhaps not stretching things too far to regard the two foregoing theses as illustrating and exemplifying (to whatever extent DS-systems of the appropriate kind are actually realized as a physical reality) Aristotle's concept that there can be states of affairs (the so-called "accidents"), whose occurrence lies outside the reach of scientific explanation — which for him is based upon principles regarding that which occurs either "always" or "for the most part." The foregoing examples teach a yet more significant lesson. For despite the fact that, in the systems considered, we have a *complete knowledge* of the functioning of the system (in that we know the set of possible states and all "laws of transition" governing the succession of these states) we can neither predict nor retrodict nor even explain certain occurrences.[10] Here scientific understanding can, it would appear on the basis of our examples, coexist with an impotence to explain (even in principle) certain particular occurrences. This line of thought both accords with and bears out (insofar as correct) Aristotle's conception that the fundamental task of science is to concern itself with laws, with the general (i.e., with "that which happens always or usually," as Aristotle puts it) and not with the *idiosyncratic* particularities of individual cases (i.e., with the so-called "accidents").

However, the state of affairs envisaged in our theses (T5.1) and (T5.2) is by no means necessary, as is shown by the thesis

(T5.3)   *Even in an indeterministic DS-system, D-explanation may be uniformly (i.e., for all states) possible (despite the impossibility of uniform D-prediction to be noted in the next thesis (T5.4)).*

DISCUSSION.   Consider the system

	S_1	S_2	S_3
S_1	0	.5	.5
S_2	0	0	1
S_3	1	0	0

10. We are not here maintaining the *ontological* thesis that systems of this sort are common in nature, but the *epistemological* thesis that they are possible, and that consequently the prospects and possibilities to which they give rise can be dismissed (if at all) only on empirical grounds, and not on the basis of *a priori* or purely methodological considerations.

Note that to D-explain the state of the system at any time t —
whatever this may be — we need only know (at most) the immediately preceding state $st(t-1)$ and the immediately succeed-
ing state $st(t+1)$. (This is unaffected by the fact that we cannot
even P_w-predict from S_1.)

The preceding example illustrates an interesting circumstance.
A state of affairs may be characterized as "*a parte post*" explicable when *some* information about its subsequent states is
required for its explicability (in whatever mode of explanation is
at issue). To illustrate the workings of this idea let it be supposed
that we are asked to explain the occurrence of S_2 at time t. No
matter how much information we may be given about earlier
states of the system we will not be able to explain (even P_w)
S_2-at-t. But if we are told that S_3-at-$(t+1)$ then, knowing S_2-at-
$(t-1)$ we can explain, and indeed D-explain, the occurrence of
S_2-at-t. This occurrence is thus *a parte post* explicable in the
sense of this concept that has just been introduced.

Turning now to prediction, we establish first the thesis:

(T5.4)   *While D-prediction cannot, in an indeterministic DS-*
system, be uniformly (i.e., for all states) possible,
P_s-*prediction and a* fortiori P_w-*prediction, may be*
uniformly possible.

DISCUSSION. The first part of our thesis is self-evident
from the very nature of "indeterministic" DS-systems. The
second part of the thesis is shown by the example of the system

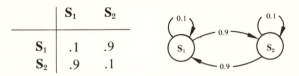

	S_1	S_2
S_1	.1	.9
S_2	.9	.1

It is evident that, given $st(t)$, we can always make a P_s-prediction
of $st(t+1)$ as being the "other" state.

The object of *retrodiction* is to "reconstruct" the past. Given
the laws of the system and certain of its states we want to
determine the temporally prior ones. If we contemplate the

possibility of a trace-annihilating system, where we never have any noninferential knowledge of the past, the point of retrodiction comes home to us.

With respect to retrodiction, we have the thesis

(T5.5) *With regard to indeterministic DS-systems, D-retrodiction may be uniformly (i.e., for all states) possible in an* infinite *system, but cannot be uniformly possible for a finite system.*

DISCUSSION. The first part of the thesis is shown by the example of the system characterized by the transition-diagram

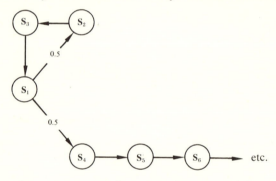

Note here that every state has exactly one arriving arrow, so that its predecessor state is always determinable.

However, in a finite indeterministic DS-system, some state must necessarily have two arriving arrows, so that D-retrodiction here becomes impossible.

The foregoing result is complemented by the thesis

(T5.6) *Even in a finite indeterministic DS-system, P_s-retrodiction may be uniformly possible.*

DISCUSSION. See the example given previously in the discussion of (T5.4).

We also have the following thesis:

(T5.7) *Even in a completely indeterministic system, D-retrodiction may for some states be possible.*

DISCUSSION. The thesis at issue is established by the example of the system

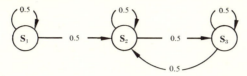

Given that the "present" state of the system is S_1, we can D-retrodict that *all* its previous states have uniformly been S_1.

Suppose we are in a position to D-retrodict the entire history of every possible state of some system. Then there can never be two arrows arriving at the same state in the transition diagram. But then if the system is finite, there can never be two arrows departing either, and so the system must (if finite) be strictly deterministic. But in an infinite DS-system, D-retrodiction of the entire history of every possible state of the system may be possible even if the system is not strictly deterministic:

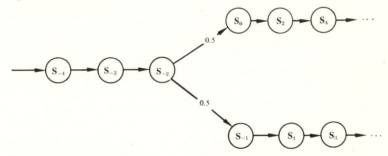

Our remaining theses will concern the relationship between explanation and prediction in DS-systems.

We begin with the thesis

(T5.8) *Even when an indeterministic DS-system is such that P_s-prediction (and a fortiori P_s-explanation) is uniformly (i.e., for all states) possible, it may yet happen that D-explanation is uniformly impossible.*

DISCUSSION. This thesis is also established by the same example cited in the previous discussion of thesis (T5.4).

Our next result is given in the thesis

(T5.9) *In an indeterministic DS-system, it may happen that P_w-prediction (and thus P_w-explanation) is uniformly (i.e., for all states) possible, whereas P_s-explanation (and thus P_s-prediction) is uniformly impossible.*

DISCUSSION. This thesis is established by the example of the system

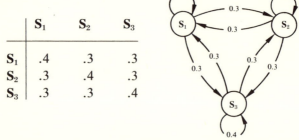

	S_1	S_2	S_3
S_1	.4	.3	.3
S_2	.3	.4	.3
S_3	.3	.3	.4

This system exhibits the interesting feature that, *if we should refuse to regard the weak mode of P_w-explanation as representing a permissible form of "explanation" proper,* then there are systems in which predictions are always warranted (uniformly — in all states of the system) but we can never "explain" anything.

A further final result is given in the thesis

(T5.10) *In an indeterministic DS-system, it may happen that D-explanation is uniformly (i.e., for all states) possible although neither P_w-prediction (let alone D-prediction) nor P_w-retrodiction (let alone D-retrodiction) are uniformly possible.*

DISCUSSION. Our result is established by the example

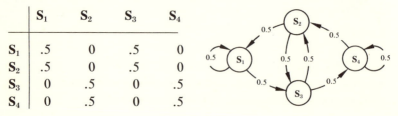

	S_1	S_2	S_3	S_4
S_1	.5	0	.5	0
S_2	.5	0	.5	0
S_3	0	.5	0	.5
S_4	0	.5	0	.5

Note that here, while we can never even weakly predict or retrodict, that given *both* the successor and predecessor of an "unknown" state, we can always deduce this unknown state itself.

In summarizing, the bearing of theses (T5.4) to (T5.10) upon the theory of explanation, prediction, and retrodiction can be put as follows — that rather than applying to indeterministic systems in any symmetric or parallel way, these resources of scientific understanding can be regarded as being virtually independent of one another in their applicability, except for the fact that when-ever prediction or retrodiction (of a given type) is possible, so then is explanation (of that corresponding type).

* * *

We will conclude this section with a series of remarks regard-ing the predictability of DS-systems.

An indeterministic DS-system can be such that its near-term future is predictable (even D-predictable), but its distant future is not even P_w-predictable. For, consider the system

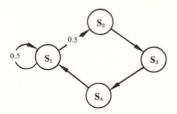

Assume that the "present" state of the system is $st(i) = S_2$. Note that we can D-predict that $st(i+1) = S_3$, $st(i+2) = S_4$, and $st(i+3) = S_1$. But we now cannot even P_w-predict $st(i+3+j)$ for any $j \neq 0$.

Contrariwise, an indeterministic DS-system can be such that its near-term future cannot be predicted (not even P_w-predicted), whereas its long-term future can be predicted (even D-predicted). This is demonstrable by the system

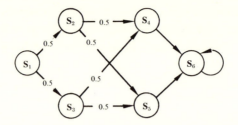

Assume that the "present" state of the system is given as $st(i) =$ S_1. We now cannot even P_w-predict $st(i+1)$ and $st(i+2)$. But we can D-predict that $st(i+3) = S_6$, and likewise all subsequent states (due to the character of S_6 as an "absorption-barrier").

Further, it is an interesting feature of systems of the sort we are studying that P_s-predictability (and also P_w-predictability) need not be transitive, if the "probable" successor of S_1 is S_2, and of S_2 is S_3, then it need not be that the "probable" 2-period successor of S_1 is S_3. For consider the system

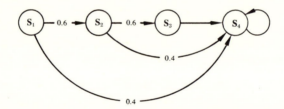

Note that although the P_s-predictable successor of S_1 is S_2, and of S_2 is S_3, we must P_s-predict S_4 as the 2-stage successor of S_1.

Finally, several interesting facts are illustrated by the system-type

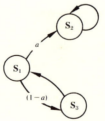

Assuming that the "present" state of the system is $st(i) = \mathbf{S}_1$, let us define $X(j)$ as the probability that $st(i+j) = \mathbf{S}_2$. It is readily seen that $s(j)$ takes on the values shown in the tabulation:

j	$X(j)$
1, 2	a
3, 4	$a + (1-a) \cdot a$
5, 6	$a + (1-a) \cdot a + (1-a)^2 \cdot a$
$2n+1, 2n+2$	$a[(n-a)^0 + (1-a)^1 + \cdots + (1-a)^n]$

But whenever a is greater than zero, i.e., whenever $1-a$ is less than one, we have it that the sum,

$$(1-a)^1 + (1-a)^2 + \cdots + (1-a)^n$$

is equal to

$$\frac{(1-a)[1-(1-a)^n]}{1-(1-a)} = \frac{(1-a)[1-(1-a)^n]}{a}$$

As a consequence, both $X(2n+1)$ and $X(2n+2)$ are equal to

$$a + (1-a)[1-(1-a)^n]$$

Now it is clear that, by taking n to be sufficiently large, we can bring this quantity as close as we please to

$$a + (1-a)(1-0) = 1$$

Thus no matter how small $X(j)$ may be for small values of j, there must be *some* value of j, say k, such that $X(j)$ invariably assumes values greater than 0.5, whenever we have $j \geqslant k$. These considerations point toward two facts:

(i) The particular fact that, in the specific DS-system considered above, no matter how small a may be, there will always be some point *up to which* the most probable multi-interval successor of $\mathbf{S}_1$ is uniformly $\mathbf{S}_3$ (assuming of course that $a < 0.5$), and *after* which it is uniformly $\mathbf{S}_2$. Furthermore, by taking a to be sufficiently small, we can put this "switching point" as far off into the future as we please.

(ii) The example also illustrates the general fact that whenever a finite indeterministic DS-system has exactly one
"absorption-barrier" state (like S_2 in the example),
then no matter how unlikely an initial entry into this state
may be, this state will *ultimately* always be the most
likely (distant) successor of any given state, provided
only that there is no closed cycle within the system of
such a kind that access to the "absorption-barrier"
state is inaccessible from it — as with the cycle beginning
at S_3 in the system

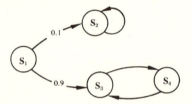

An absorption-barrier system of this general sort can serve to
illustrate an interesting aspect of the relative importance of information regarding the other states of the system in making
inferences as to an unknown state. If the system is of the type

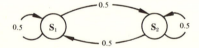

then information about its other states is simply useless for
determining an unknown state. On the other hand, if the system
has an overwhelmingly probable absorption-barrier state of the
type

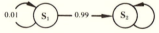

then specific information about its states is virtually unnecessary.
So long as we know it *has* a past history at all, any unknown state
will be S_2 with an overwhelming probability. These examples
illustrate the fact that while knowledge of a system's states alone,

wholly unmediated by any knowledge — at least implicit knowledge — of its laws, can never be a sufficient basis for a reliable inference regarding an unknown state, knowledge of its laws can at least in some cases serve this purpose even in the face of a complete absence of information about its states.

6. Infinite Discrete State Systems

It is also of considerable interest to devote more explicit attention than has yet been given to *infinite* discrete state systems. Such systems will (by definition) continue to exhibit the feature of temporal discreteness, so that their history can also be represented as a (conceivably doubly infinite) series:

$$\ldots, st(t_0-2), st(t_0-1), st(t_0), st(t_0+1), st(t_0+2), \ldots$$

But now the situation is no longer that $st(t)$ must be one of a restricted group of states:

$st(t) =$ one element of the finite list of alternatives: $S_1, S_2, \ldots, S_n$

Rather, $st(t)$ can now possess an infinite variety:

$st(t) =$ one element of an infinite set of alternatives

An interesting example arises when we consider a system whose "state" is an (initially infinite) string of 0's and 1's:

$$\ldots 01001$$

Let the law of transition be that if $s(t)$ is the sequence $\ldots xyz$, then

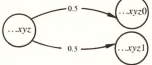

Note that this system will illustrate the thesis

(T6.1) *There are infinite probabilistic DS-systems such that given the state of the system at any time t whatsoever,*

> *one can always D-retrodict the entire infinite prior history of the system but we can never even P_w-predict its state at $t + 1$, nor* (a fortiori) *at any future time.*

An especially significant application of this machinery of infinite discrete state systems will result whenever the state of a system at any time t is given by one or more *state-functions* of a (continuous) universal parameter, say

$$st(t) = \big(f(t), g(t), h(t), \ldots\big)$$

where $f(t)$, $g(t)$, $h(t)$ are numerical functions of the continuous time parameter t. Examples would be provided by such physical parameters as temperature, mass, velocity, etc., that have infinite variability.

A particularly interesting case arises when we set out from the starting point of a finite DS-system, but now define a "state" of this system as its *second-order probabilistic state*, to be construed as a probability distribution over its ordinary, first-order basic or primary states of the orthodox type. This transforms the initially finite DS-system into an infinite one. An example will clarify matters. Consider the finite (indeterministic) DS-system

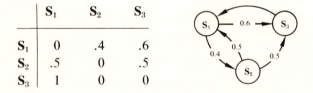

	S_1	S_2	S_3
S_1	0	.4	.6
S_2	.5	0	.5
S_3	1	0	0

Now assume that an initial second-order (probabilistic) state of the system is $st(t) = (x_1, x_2, x_3)$ where $x_i = $ the probability that the system is in basic state S_i (clearly we will have it that $x_1 + x_2 + x_3 = 1$). Then its next state is obviously

$$st(t + 1) = s(t) \times M = (x_1, x_2, x_3) \times M$$

where M is the transition matrix at issue. Thus

$$st(t + 1) = (0.5x_2 + x_3, 0.4x_1, 0.6x_1 + 0.5x_2)$$

Thus, for example, if $st(t_0) = (\frac{1}{3}, \frac{1}{3}, \frac{1}{3})$, then

$$st(t_0 + 1) = (\tfrac{15}{30}, \tfrac{4}{30}, \tfrac{11}{30})$$
$$st(t_0 + 2) = (\tfrac{13}{30}, \tfrac{6}{30}, \tfrac{11}{30})$$
$$st(t_0 + 3) = (\tfrac{35}{75}, \tfrac{13}{75}, \tfrac{27}{75})$$

This process is readily pushed as far as we please, and will in general yield a perfectly definite probabilistic metastate for any future juncture we care to investigate. Of course the three initially possible states now give rise to an infinite variety of probabilistic "states."

This example serves to illustrate an interesting and important general fact:

(T6.2) *An indeterministic finite DS-system can invariably be portrayed as a strictly deterministic infinite state system by this device of letting a (derived) state of the new system be a probability-distribution across the (fundamental) states of the initial system.*

7. The Comparative Evaluation of Probabilistic Explanations[11]

It is a fact well known from deductive logic that one and the same conclusion can be established by a multiplicity of distinct and quite different arguments. A simple example of this is afforded by the two syllogisms

All A is B	No D is A
Some C is not B	Some C is D
Some C is not A	Some C is not A

An exactly parallel situation obtains with respect to explanatory arguments: Very different explanations (explananses) can lead to

11. This section is a slightly altered version of a joint paper with F. Brian Skyrms published under the title "A Methodological Problem in the Evaluation of Explanations" in *Nous*, vol. 2 (1968), pp. 121–129. This material is used here with the kind permission of Dr. Skyrms and of the Wayne State University Press.

exactly the same explanatory conclusion (explanandum). Thus if a wire breaks under the pressure of a given weight, this might be because the imposed tension was too high for its tensile strength, or because the wire was defective, or even both. All three of these would be "*possible* explanations" of the fact at issue (the wire's breaking). It is critically important for the theory of explanation to consider the ways in which the relative merits of such alternative explanations are to be assessed. It will be shown here that the idea of the "goodness" of an explanation actually involves two significantly distinct conceptions, with the aid of the conditional probability model to be developed.

Some Useful Machinery. Consider the following situation. X tosses a coin supplied by Y. Y calls heads and the coin comes up heads. X requests an explanation of this occurrence from a bystander. The bystander knows that recently a novelty shop in the area has begun selling coins biased $7:3$ in favor of heads. Two explanatory accounts thus suggest themselves to him: (E_1) that the coin was fair (and so happened to come up heads), and (E_2) that the coin was biased $7:3$ (and so happened to come up heads). Each of these alternative explanatory accounts defines a probability distribution over the set of all possible occurrences $(S_1 = \text{heads}; \quad S_2 = \text{tails})$. The bystander's information about these probabilities may thus be tabulated in a matrix, $// Pr(S_i/E_j)//$, comprising the conditional probabilities of the various possible occurrences relative to the several explanatory accounts. The specific matrix for the given example would be

$Pr(S_i/E_j)$	E_1	E_2
→ S_1: Heads	.5	.7
S_2: Tails	.5	.3

The arrow indicates that it is S_1 which has actually occurred and is to be explained. Note that, since the S_i are an exhaustive set of mutually exclusively elements, the columns of the matrix must sum to one.

In general, we suppose that the subject of discussion is a physical system of some sort which is capable of exhibiting, at any given time, one or another of some group of states $S_1, S_2, \ldots,$ S_n; where these S_i are assumed to be mutually exclusive and exhaustive. We suppose also that a number of distinct explanatory accounts $E_1, E_2, \ldots, E_m$ are at issue. We assume further that some mechanism for dealing with the relevant conditional probabilities is at our disposal, so that in each case a conditional probability matrix of the type $//Pr(S_i/E_j)//$ is defined.

In the above example, the explanatory accounts were themselves mutually exclusive. However, our discussion will not require us to make this or any other *general* assumption regarding the nature of the explanatory accounts at issue.

It would be quite incorrect to claim that the two criteria which we are about to distinguish are the *only* relevant factors in the assessment of an explanation. Therefore, it will help the reader to focus on the distinctions we make if throughout this section he assumes that the explanations in question are on a par with respect to all *other* criteria of explanatory goodness. As examples of such other criteria we might mention the *confirmation* of the explanatory accounts, and their *relevance* to the occurrence to be explained. Thus, for example, suppose we wish to provide an explanation for the fact that $S_1 = Mr.$ *X (a rug dealer in Cleveland, Ohio) is of Armenian descent.* Consider now the following three explanatory accounts:

E_1: Mr. X is an American rug dealer, and 90 per cent of them are Armenian.

E_2: Mr. X is an Ohio rug dealer, and 70 per cent of them are Armenian.

E_3: Mr. X is a Cleveland rug dealer, and 30 per cent of them are Armenian.

We obtain the conditional probability matrix

$Pr(\mathbf{S}_i/\mathbf{E}_j)$	$\mathbf{E}_1$	$\mathbf{E}_2$	$\mathbf{E}_3$
$\longrightarrow \mathbf{S}_1$	.9	.7	.3
$\mathbf{S}_2 = \sim S_1$	.1	.3	.7

Although $\mathbf{E}_3$ is the "worst" explanation of $\mathbf{S}_1$ of the group $\mathbf{E}_1 - \mathbf{E}_3$ because it endows $\mathbf{S}_1$ with the lower conditional probability than any of its competitors do, one is not in a position to discard $\mathbf{E}_3$ in favor of $\mathbf{E}_1$ or $\mathbf{E}_2$ because of its greater *relevance* to the item(s) at issue in $\mathbf{S}_1$.

Brief notice must be taken of the question of choice in the specification of a set of "states." In some systems, one particular set of states is so much at the forefront of our conception of the system that this strikes us as essentially the only possible way to sort the behavior of the system into states (e.g., the foregoing coin-toss illustration). In other cases, the process of "slicing the pie" of possible circumstances into an exhaustive set of mutually exclusive alternative states can be an essentially arbitrary undertaking. Thus suppose that two coins, a and b, are tossed simultaneously. One acceptable possibility set of outcome states would be $(\mathbf{S}_1)\ \mathbf{H}_a\mathbf{H}_b$; $(\mathbf{S}_2)\ \mathbf{H}_a\mathbf{T}_b$; $(\mathbf{S}_3)\ \mathbf{T}_a\mathbf{H}_b$; $(\mathbf{S}_4)\ \mathbf{T}_a\mathbf{T}_b$. Another equally acceptable possibility set would be : $\mathbf{S}_1 = $ two heads; $\mathbf{S}_2 = $ one head and one tail; $\mathbf{S}_3 = $ two tails. In general, it should be noted that an acceptable possibility set can always be devised by taking $\mathbf{S}_1$ to be some particular state of the system, and then taking $\mathbf{S}_2$ as the state of not being in $\mathbf{S}_1$ (for short, $\mathbf{S}_2 = \mathbf{S}_1$). However, the admitted possibility of choice in partitioning the behavior of a system into "states" does *not* constitute any insuperable difficulty for our approach to the study of explanations. The particular partitioning of states at issue when any *concrete* question of evaluating an explanation arises, is generally dictated by (i) the dialectical context within which the explanations are to be evaluated and/or (ii) the presence of a common theoretical framework presupposed by all the available explanatory accounts. Usually the concrete context of discussion removes the arbitrariness which could, abstractly speaking, be present.

The problem we propose to consider can now be formulated in a very general and precise way as follows. How, given a conditional probability matrix of the stipulated kind, is one to formulate—in terms of the information provided by this matrix— suitable criteria for assessing the "goodness" of explanations?

To start with, consider the following example. We are seeking to explain the actual circumstance that some item a has the property P (or belongs to the class of things having P, i.e., $a \in P$). We are offered three explanatory accounts.

E_1: $(a \in Q)$ & (Almost all Q's are P's)
E_2: $(a \in R)$ & (A bare majority of R's are P's)
E_3: $(a \in T)$ & (Many, but only a minority of T's are P's)

It is intuitively clear that E_2 is a better explanation of the actual state than E_3, and that E_1 is a better explanation than E_2 or E_3. Thus in one sense, the "goodness" of an explanation is its *comparative strength* as determined by the degree to which, as contrasted with other explanations, it confers greater likelihood upon the state to be explained. Comparative strength is easily seen to be determined by *the extent of dominance in the row (of the conditional probability matrix) corresponding to the state of affairs to be explained*.

The foregoing example, for instance, would yield a matrix of conditional probabilities somewhat as follows.

	E_1	E_2	E_3
⟶ S_1: $(a \in P)$	.9	.6	.3
S_2: $(a \notin P)$	.1	.4	.7

It is clear at a glance that E_1, which accords the largest probability value to S_1, has the greatest comparative strength of any of the three explanations in question.

In general, if S_k is the actual state to be explained, and if $Pr(S_k/E_i) > Pr(S_k/E_j)$ then E_i will be said to be a *stronger* explanation of S_k than E_j. That explanation (if any) among the E_i for which $Pr(S_k/E_i)$ is a maximum, is to be characterized as the *strongest available explanation* for S_k. An inspection of the sample matrix given above will show that, in such cases, the formal definitions preserve our intuitive value ordering.

This purely ordinal formalization of the idea of comparative strength may be supplemented by a measure of the relative *degree* to which one explanation is stronger than another. The most obvious measure that suggests itself is the numerical difference between the conditional probabilities of the explanations of the state to be explained. Thus in our example, E_1 would be stronger than E_2 by 0.3 (and E_2 would be stronger than E_1 by -0.3, i.e., weaker by 0.3). The difference in strength between two explanatory accounts must thus always be between -1 and $+1$. On this approach, we may define a measure of the comparative strength (within a range of several available explanations) of the strongest available explanation E_k of a given state S_m to be

$$\frac{1}{n-1} \sum_{i=1}^{n} [Pr(S_m/E_k) - Pr(S_m/E_i)]$$

where n is the number of explanatory accounts at issue. It is clear that this quantity will range between 0 and 1.

To this point the "goodness" of an explanation (relative to other available explanations) has been discussed solely with reference to the row relations within our conditional probability model. There is, however, another, more absolute sense of the "goodness" of an explanation, to wit, the extent to which it singles out the (actual) state to be explained from the set of possible alternative states. Let us call this sense of the "goodness" of an explanation its *explanatory power*. Explanatory power will have to be measured with reference to the column relationships of the conditional probability matrix, and will be determined by *the extent to which the position corresponding to the state of*

*affairs to be explained has dominance in the column (of the con-
ditional probability matrix).*

Thus consider the matrix

	E_1	E_2	E_3
$\longrightarrow S_1$	1	.7	.4
S_2	0	.3	.3
S_3	0	0	.3

The three explanations exemplify three importantly different levels of explanatory power.

(I) E_1 confers probability 1 on the state to be explained; it thus *renders all the other (mutually exclusive) states "effectively impossible"* (i.e., of probability 0). In general, if one of the column entries is 1, then all the other entries in that column must be zeros (since the entries of any given column must sum to 1). The condition that $Pr(S_k/E_i) = 1$ therefore corresponds to the requirement that E_i must single out the given state S_k (from the set of alternative states) *deductively* (at any rate provided that we are prepared to construe "effective" as genuine impossibility). An explanation which satisfies this condition has the greatest possible explanatory power.

(II) E_2 satisfies the weaker condition that $Pr(S_k/E_i) > 0.5$. Since the columns must sum to one, this condition guarantees that under the hypothesis E_1, S_k will be *more likely than all the other states taken together.* In other words, given E_i, S_k will be more likely than not.

(III) E_3 satisfies the still weaker condition that $Pr(S_k/E_i)$ simply be the *largest* entry in its column. Fulfillment of this condition guarantees that under the hypothesis E_i, S_k must be *the most likely single state.*

Clearly, an explanation which fulfills condition (I) has greater explanatory power than one which fulfills only conditions (II) and (III); one which fulfills condition (II) has greater power than one which fulfills only condition (III); and one which fulfills condition (III) is more powerful than one which fulfills neither (I), (II) nor (III). And this is so not only in the context of our technical sense of "explanatory power," but also with our informal understanding of these matters.

If our explanatory account fulfills a more stringent condition than some competitors, as regards explanatory power in the sequence (I), (II), (III), we shall call it a *more powerful* explanation. Explanatory power is clearly yet another determinant of the "goodness" of explanations.

It remains to consider what formal relationships, if any, hold between the comparative strength of explanations and their explanatory power. The reader may already have noticed that in the last matrix given, the ordering of the explanatory accounts E_i on the basis of comparative strength *coincides* with the ordering on the basis of explanatory power. It is an interesting fact that such agreement need by no means be the case in general. Thus consider the following conditional-probability matrix:

	E_1	E_2
S_1	0	.1
$\longrightarrow S_2$	.4	.3
S_3	.6	.2
S_4	0	.2
S_5	0	.2

Although E_1 is the strongest available explanation of S_2 as the actual state, E_2 is the more powerful explanation of this occurrence. Here then we see, in effect, a conflict between these two determinants of explanatory "goodness."

A word must be said about the circumstances under which such discord between the "explanatory power" and the "comparative strength" of explanations can arise. This can only happen when the most powerful available explanation fulfills *only* condition (III) of the preceding listing. It is easy to see that whenever the most powerful explanation fulfills either one of conditions (I) or (II), then it must also necessarily be the strongest available explanation. The possibility of a discord between our two criteria of the "goodness" of explanations is thus localized, being restricted to the case of explanations that are very "weak explanations" (in terms of their explanatory power). For this reason, the present considerations regarding the evaluation of explanations are relevant only to probabilistic and not to deductive explanations. (It is also easily shown that if we consider the matrix for only two states, namely (i) the state to be explained S_1, and (ii) the state of not being in $S_1 (S_2 = \sim S_1)$, a conflict cannot occur. Thus the set of explanatory alternatives within which the given actual state is being explained constitutes a crucial factor.)

The fact that this general agreement between the two specified determinants of explanatory "goodness" must necessarily obtain with all reasonably powerful explanations is doubtless part of the reason why these criteria for the assessment of explanations have not hitherto been discriminated with sufficient care. Unquestionably another important reason is the almost exclusive focus of methodological discussions, up to the most recent times, upon *deductive explanations*, to the exclusion of explanations in which a significant role is played by statistical and probabilistic considerations.

To summarize this discussion: We have distinguished two factors which are relevant to the assessment of the "goodness" of an explanation

(i) Its *comparative strength* as determined by the extent to which it renders the occurrence to be explained *more likely than other alternative explanations* manage to do, and

(ii) Its *explanatory power* as determined by the extent to which it renders the occurrence to be explained *more likely than other alternative occurrences.*

Introducing the machinery of a conditional probability matrix to register the probabilities of the possible alternative occurrences relative to the several given explanatory accounts, we see that these two factors are, viewed conceptually, sharply distinct: the former having to do with row relationships in the conditional probability matrix; the latter with column relationships. The discussion has shown that there will in some cases be an actual discord between these two criteria. In practice, an outright conflict between the explanatory power and the comparative strength of explanations can only occur in cases of explanations with low explanatory power. This localization of the area of possible discord should not, however, be taken as militating against the conceptual importance of the distinctness of these two criteria.

8. Chronologically Teleological Explanation and Teleology versus Mechanism

In the present section the leading conceptions of the foregoing discussion of explanatory concepts will be brought to bear upon the much agitated issue of "teleology" vs "mechanism" in the theory of explanation. But first some preliminary considerations regarding the essential ideas of teleological and mechanistic explanation are in order.

For purposes of the present discussion, the concept of "teleology" (and its opposite) will be construed in a somewhat artificial and technical chronologized sense. In Aristotle, a founding father of the teleological school of thought, teleology has to do fundamentally with explanations given in terms of functions and purposes. In Leibniz, the principal telelogist of modern philosophy, teleology has to do with "economy" or "simplicity" or "efficiency" in nature as evidenced in the basic role of minimax and conservation principles in physical science. For our present purposes, however, we lay aside these classical concepts of teleology. We shall here construe "teleology" in a way that relates solely to the role played by the time factor in explanation, as evidenced in the temporal scope of the data requisite for explanatory purposes. On this approach, the issue of teleological vs mechanistic explanation comes down to the difference between the process of *a fronte* explanation of events in terms of their chronologically posterior successors as contrasted with *a tergo* explanation of events in terms of their chronologically anterior predecessors.[12] On this basis then, a "mechanist" will insist on the general sufficiency of *a fronte* explanation in science whereas a "teleologist" will maintain the necessity of *a tergo* explanations.

From this standpoint of the chronology of explanatory data, there will be no one monolithic and unitary concept of teleology, but rather a whole spectrum of teleological positions. Specifically, room will have to be made for at least six major ways of formulating the doctrine of "teleology":

(T1) *Every occurrence can be explained* a fronte *only (so that no occurrence can possibly be explained* a tergo*).*

(T2) *Every occurrence can be explained* a fronte *(although some—or even all—occurrences can possibly also be explained* a tergo*).*

(T3) *Every occurrence can be explained only if some* a fronte *(i.e., chronologically posterior) data are available (and cannot possibly be explained solely* a tergo*; although some* a tergo *(i.e., chronologically*

12. One recent precedent for this specifically chronological construction of teleology is afforded by the discussion of Section 4 of A. Grünbaum, "Temporally Asymmetric Principles, Parity between Explanation and Prediction, and Mechanism versus Teleology," *Philosophy of Science*, vol. 29 (1962), pp. 146–170.

> *anterior) data may actually be needed for its explana-*
> *tion in some — or even in all — cases).*

(T4)–(T6)   *Exactly as with (T1)–(T3) above, except for chang-*
             ing the initial "every" to "some."

This inventory of six formulations of "teleology" lists the positions in order of what may reasonably be characterized as decreasing strength. On this basis we can speak of comparatively "weaker" and "stronger" formulations of teleology. Furthermore, each thesis is itself equivocal with respect to its inclusion of the key term "explained," which can be construed in either a stronger or a weaker sense (D-explanation or P-explanation).

Let the term "explanation" of (T1)–(T6) be for the present construed in the sense of D-explanation, and let us now raise the question of the extent to which these "teleological" positions can be associated with flesh-and-blood adherents. To the best of my knowledge, no philosopher, scientist, or theoretician has ever espoused (T1). This is certainly rather a showcase version of teleology, whose interest is rather speculative than historical. On the other hand, there is no question that Leibniz espoused (T2), and there is some possibility (though I have reservations on this head) that Aristotle may be regarded as a subscriber to (T3). Thus (T2) and (T3) are the "strongest" versions of "teleology" (in our sense of this term) that can lay claim to historically legitimating credentials.

Corresponding to the six foregoing versions of "teleology" there are an equal number of counterpart versions of "mechanism," its contradictory, obtained in each case by the simple expedient of *interchanging* the expression "*a fronte*" and "*a tergo*" in the previous formulations of (T1)–(T6). The six resultant versions of "mechanism" are as follows:

(M1)   *Every occurrence can be explained* a tergo *only (so that no occurrence can possibly be explained* a fronte).

(M2) *Every occurrence can be explained a* tergo *(although some, or even all, occurrences can possibly also be explained a* fronte*).*

(M3) *Every occurrence can be explained only if some a* tergo *(i.e., chronologically anterior) data are available; and cannot possibly be explained a* fronte, *although some a* fronte *(i.e., chronologically posterior) data may actually be needed for its explanation in some, or even in all, cases.*

(M4)–(M6) *Exactly as with (M1)–(M3) above, except for changing the initial "every" to "some."*

Historically speaking, there are but few (M1)-mechanists, of whom perhaps the most notable is Descartes. But the philosophical woods are full of (M2)-mechanists, including of course Leibniz, who argues at much length the compatibility of (M2) with (T2).

It is important to recognize the paucity of contradiction relations between the various kinds of "mechanism" and "teleology." They are exhausted by the following tabulation of 15 out of 36 possibilities:

    (i) (T1) is incompatible with all (Mi).

    (ii) (M1) is incompatible with all (Ti).

    (iii) (M2) is incompatible with (T3) and (T5).

    (iv) (T2) is incompatible with (M3) and (M5).

This listing of various contradiction relations that obtain between the several versions of chronological mechanism and teleology is complete. In consequence all sorts of concurrent and simultaneous espousals of versions of both mechanism and teleology are conceptually possible. Since (T1) is effectively a "strawman" thesis which has never found actual advocacy, we may view the conflict of "teleology" vs "mechanism" as exemplified by:

    (I) The conflict between (M1) and the various (Ti), and particularly (T2) as the "strongest" version of teleology. (Incidentally, from *this* standpoint it would appear that it is "teleology," and not "mechanism" that merits the palm of victory, since (M1) finds little warrant or support on the contemporary scientific scene.)

(II) The conflict between (M2), on the one hand, and (T3) or (T5) on the other.

(III) The conflict between (M3) or (M5), on the one hand, and (T2) on the other.

We may, I believe, justifiably regard (II) and (III) as the only "realistic" versions of the "conflict of mechanism vs teleology." But in any event, any accurate discussion of this problem requires fine discriminations to be made between various *types* of "mechanism" and "teleology," discriminations which may, as our discussion has shown, be far more subtle than at first meets the eye.

Returning to the mainstream of our discussion, we shall now illustrate concretely by means of DS-systems how some of the various versions of mechanism and teleology can coexist in physical systems.

Illustrative Case 1. A DS-system which exemplifies (T2) and (M2), with "explanation" understood as D-explanation:

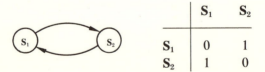

	S_1	S_2
S_1	0	1
S_2	1	0

DISCUSSION. It is at once evident that, given the actual state of the system at any time $t = t_0$, the entire history of the system, past and future, can be specified.

Illustrative Case 2. A DS-system (infinite) which exemplifies (T1) — thus precluding all (Mi) — with "explanation" understood as D-explanation.

and in general

DISCUSSION. Note that each S_i has only one arrow arriving (so that the past history of all previous states can be reconstructed once the current state is given), but that every S_i has two equiprobable departing arrows, so that the future states can never be predicted (in any sense of "prediction"). And consequently, a given state can never be "explained" (in any of the three senses) in terms of data regarding only antecedent states.

Illustrative Case 3. A DS-system which exemplifies (T3) — but not (T2) — and also exemplifies (M3) — but not (M2) — with "explanation" understood as D-explanation.

DISCUSSION. Consider again the example cited in the discussion of thesis (T5. 10). Note that only if we know both *the immediate predecessor and successor* of a given state can we explain (in the deductive sense) that state itself. That is, only if we know both X and Y can we (deductively) fill in the blank in the sequence $X - Y$.

Of course DS-systems can also readily be exhibited to illustrate the other logically possible coexistence relations of the various types of "mechanism" and "teleology." The principal reason for selecting the foregoing three cases is their capacity to illustrate concretely the possibility of systems exemplifying the three strongest types of "teleology."

One important point regarding "teleology" (in our narrower sense) is clearly brought out by the foregoing considerations. This is the fact that there is nothing improper and illegitimate about teleology *per se*, at any rate from the standpoint of the special kind of "teleology" under consideration here. For it is a strictly empirical question whether a physical system has laws of the type that render it somehow "teleological" in the sense of our discussion. Thus there is no justification for any broadside dismissal of teleology as intrinsically unscientific. And indeed it must be recognized that even a system that is "mechanistic" in some appropriate sense may yet be "teleological" in another.

It must be stressed that the foregoing considerations are of an epistemological and not ontological character. We have made no attempt here to argue the thesis (which may be true, but is beside the point in any case) that teleogical and/or mechanistic systems of the types described are commonplace in physical realization, being frequently exemplified in nature. The main aim has been to show that the issue of mechanism vs teleology, when viewed after the chronological manner of our present discussion, is an essentially empirical, scientific question that waits upon a study of the actual facts, and is not a theoretical issue that can be settled by abstract reasoning from *a priori* premises or by any decree or fiat on the basis of methodological or of conceptual analysis.

9. Explanation and Its Cognates as Fundamentally Evidential Reasonings: Evidence versus Demonstration

The epistemological concept in terms of which explanation and its cognates, prediction and retrodiction, can most effectively be explicated is the broader concept of supporting evidence. Here we can find a generic home for all three of these procedures: They are all particular species of the generic family of evidential reasonings. Before developing this theme, however, it is first necessary to take note of some points relating to the nature of evidence.

The question of the epistemological character of the evidential relationship—the relation obtaining between evidence statements upon the one hand and a statement supported by them upon the other—is complex and ramified. The evidence concept covers a wide variety of distinguishable species: It includes all of the special relationships that obtain when a body of discourse "supports" some proposition in any of the numerous appropriate senses of this term, ranging from the most demanding species of evidence which calls for establishment of a conclusion "beyond the shadow of a doubt," to the most provisional and tentative modes of argument, such as analogy or "circumstantial"evidence. The evidential relation holds whenever we must give some weight or credence to the conclusion upon some given statements (the evidence) as hypothesis.

The logical nature of the concept of evidence is a matter greatly in need of theoretical analysis and clarification. A study of this relationship, and an endeavor to supplant the purely qualitative concept of "constituting supporting evidence" by a qualitative measure of degree of evidential support, is one of the principal tasks of modern formal inductive logic. In recent years, important progress toward a solution of this problem has been made, particularly by K. R. Popper, C. G. Hempel, and Rudolf Carnap, and the work has also been carried forward by J. G. Kemeny and Paul Oppenheim, and others.[13] In the following section we shall enter upon this large and technical subject. For the present, it will suffice to note in a preliminary way some few characteristics of the concept of supporting evidence that emerge from the logical analysis of this concept.

13. The pioneering works are Popper's *Logik der Forschung* (Vienna, 1935); Hempel's "Studies in the Logic of Conformation," *Mind*, vol. 54 (1945), pp. 1–26, 97–121; and Carnap's *Logical Foundations of Probability* (Chicago, 1950). Other contributions include Kemeny and Oppenheim, "Degree of Factual Support," *Philosophy of Science*, vol. 19 (1952), pp. 307–342; Popper, "Degree of Confirmation," *The British Journal for the Philosophy of Science*, vol. 5 (1952), pp. 143–149; a review of the preceding by Kemeny in *The Journal of Symbolic Logic*, vol. 20 (1955), pp. 304–305; Popper, "A Second Note on Degree of Confirmation," *The British Journal for the Philosophy of Science*, vol. 7 (1957), pp. 350–353; and the writer's paper, "A Theory of Evidence," *Philosophy of Science*, vol. 25 (1958), pp. 83–94. The literature is helpfully surveyed in Henry E. Kyburg, Jr., "Recent Work in Inductive Logic," *American Philosophical Quarterly*, vol. 1 (1964), pp. 249–287.

There is, to be sure, the very significant mode of *conclusive* evidence, evidence that wholly establishes its conclusion. But by adducing evidence in support of some proposition we will not, at least in general, go so far as to *establish* this proposition. It is only necessary for evidence to render its conclusion more tenable or more likely than before, i.e., more probable *a posteriori* than *a priori*. Evidence, then, is by nature a logically weaker mode of reasoning than proof, in any of the senses of that term. The central and fundamental fact of the theory of evidence is that one statement may constitute evidence for another which goes significantly beyond it in assertive content. This at once differentiates evidence from entailment or deductive demonstration. A true statement may legitimately provide evidence for a falsehood, and one statement may constitute evidence for each of several incompatible statements.[14]

The point that deserves prime emphasis is that evidence admits a logically weaker counterpart of demonstration; it, too, is a type of justifying argumentation, but the logical relationship it establishes can be far more tenuous. (For this very reason, its range of application is much wider.) In giving a demonstration we must bring forth considerations that render the acceptance of the proposition in view well nigh mandatory. In adducing evidence we may merely bring forth considerations that render acceptance of the proposition more palatable than before. Evidence, then, is a mode of reasoning which is a logically weaker cognate of demonstration or proof. It is structurally similar in also giving reasons in support of a conclusion, but it may furnish its support in a far more inconclusive and tenuous manner.

Now it is just this distinguishing feature between evidence and demonstration that renders the concept of evidence peculiarly fitting for the logical characterization of explanations and reasoned predictions. As regards predictions, we have a reasoned

14. A detailed treatment of these, and analogous considerations will be given below.

justification of a prediction in precisely those cases in which we are confronted with the evidence relevant to the prediction, and find that this supports the predicted possibility more than its principal alternatives, i.e., when the weight of evidence in favor of the predicted eventually exceeds the weight of the evidence in favor of the competing candidates (its principal alternatives). And whatever has been said here about reasoned prediction applies to reasoned retrodiction as well. What more than this could reasonably be asked of a reasoned prediction?

But with explanation too the situation is analogous. As we have been concerned to show at some length, the weaker probabilistic modes of explanation must be recognized along with the orthodox deductive mode of explanation. (Otherwise—if only this most rigorous mode of explanation is alone recognized—then, given the stochastic revolution in modern physical science, many occurrences would have to be classed as "inexplicable" for which an account can be given that assures all possibly appropriate demands for undertaking.) And once these weaker sorts of probabilistic explanations are recognized, the fundamentally evidential nature of explanatory reasoning is exhibited. Explanations also may thus be assimilated to predictions and retrodictions as fundamentally evidential reasonings.

To summarize these considerations: It is important to distinguish between *evidential* and *demonstrative* reasonings. Explanations, prediction, and retrodiction are fundamentally evidential, and *evidence* is the epistemological concept in terms of which the kinship between these three types of empirical reasonings can most effectively be explicated. In the interest of accurate logical taxonomy it is necessary to classify all of these—explanation definitely included—as generically *evidential* rather than specifically *demonstrative* modes of reasoning. The domain of scientific explanation as a whole falls within the scope of the general theory of evidence, a field of which legal evidence, for example, would constitute yet another branch.

10.   The Logic of Evidence[15]

As was argued in the previous section, explanation as well as the cognate procedures of prediction and retrodiction are all fundamentally evidential modes of reasoning. In the interests of a fuller clarification of these concepts it is therefore desirable to examine the logic of evidence more closely. "Probable evidence," as Bishop Joseph Butler wrote in *The Analogy of Religion* (1736) "in its very nature, affords but an imperfect kind of information." We shall consider what sort of information it does afford, the better to understand in what respects it is to be considered imperfect.

In the present section, two distinct albeit related conceptions of evidence will be explicated and analyzed: *confirming evidence* by means of which a thesis is established, and *supporting evidence* which does not establish its thesis but merely renders it more tenable. The formal characteristics of each of these concepts of evidence will be examined in detail after we have dealt with the auxiliary notion of *evidential presumption*. Thereupon, these considerations will be used as basis for a survey of *rules of evidence* in order to establish the logical characteristics of the evidential relationship. Finally, we shall append some indications regarding the methodological bearing of this study of the logical theory of evidence.

The evidential relationship can be taken to obtain both between *properties* and between *statements*. Thus "being British" constitutes evidence for "speaking English," just as "Mr. X is British" constitutes evidence for "Mr. X speaks English." In the present section, for the sake of expository convenience, this duality of the evidential relationship is resolved in favor of statements. For the purposes of this inquiry, the evidential

15. A previous version of this section appeared as "A Theory of Evidence," *Philosophy of Science*, vol. 25 (1958), pp. 83–94. In writing this article the author benefited from constructive criticisms by Olaf Helmer and John G. Kemeny.

relation will be construed as a *logical relationship between state-ments*, analogous in this respect to the logical relationships of entailment, equivalence, and incompatibility. This logical relationship is to be interpreted as obtaining between two statements just in case the former affords *evidence* for the latter (in some appropriate sense).

Throughout, we shall make full use of the notion of the "probability" or the "likelihood" of a statement. A real-valued function *Pr(p)* defined for all statements *p* that belong to a body of discourse *D* is to be called a likelihood or *probability measure* on *D* if the following three conditions are satisfied:

(L1) If p is in D, then $Pr(p)$ is always a real number ≥ 0
(L2) If p is necessary, then $Pr(p) = 1$
(L3) If p and q are incompatible, then $Pr(p \lor q) = Pr(p) + Pr(q)$

Several points of usage and notation must be explained. Throughout, "p," "q," "r," ..., will be used as propositional variables with range D. The signs " $\sim$ ", " & ", and " $\lor$ " will be used for negation, conjunction, and disjunction, repectively. The arrow " $\longrightarrow$ " is used as a symbol for the logical relationship of entailment (i.e., strict, not material, implication). Further, p and q are said to be incompatible if $p \longrightarrow \sim q$. And p is *inconsistent* if incompatible with itself, and *necessary* if $\sim p$ is inconsistent.

Conditions (L1)–(L3) are axioms for the probability function, and assure that Pr obeys the usual rules for probability as developed in the mathematical theory of probability.[16] Specifically, whenever $Pr(q) \neq 0$, we can introduce a measure $Pr(p, q)$ of the *conditional likelihood* of p on q:

$$Pr(p, q) = \frac{Pr(p \& q)}{Pr(q)}$$

16. P. R. Halmos, "The Foundations of Probability," *American Mathematical Monthly*, vol. 51 (1944), pp. 493–510.

The likelihood measure *Pr* is in essentials identical with the measure introduced in connection with the definition of a *degree of confirmation* among statements. On this subject the reader is referred to R. Carnap's *Logical Foundations of Probability* (Chicago, 1950). Several methods exist for the actual computation of numerical values of such a measure for statements within languages of various (generally quite simple) types.[17] However, no details regarding any particular, specific determination of a *Pr*-measure are here required. For present purposes, the notion of a measure of statement likelihood will be used only in a general way, as an aid to facilitate precise explication of certain aspects of evidence. Indeed, the ensuing conceptual considerations regarding evidence are valid even if numerical measures of statement likelihood are wholly dispensed with. However, the use of such a measure will substantially simplify problems of explanation and exposition.

The statement *q* may be said to constitute a *presumptive factor* for the statement *p*, if *q* is well established, and if taking *q* as an hypothesis renders *p* more likely than its negation.[18] By use of the likelihood measure *Pr*, the critical condition for confirming evidence may be formulated as

$$Pr(p, q) \geqslant Pr(\sim p, q)$$

By the definition of *Pr*, this condition can be seen to amount to $Pr(p, q) \geqslant 0.5$ [or to $Pr(p \ \& \ q) \geqslant Pr(\sim p \ \& \ q)$]. Thus *q* yields an evidential presumption for *p* if *q* is a statement of such a type that, relative to it, *p* is more likely than not.

A measure of a *degree of evidential presumption* (of *p* on evidence *q*), designed to correspond to the concept embodied in

17. Carnap's *Logical Foundations of Probability* should be consulted for details and for references to the literature.

18. A presumptive factor need not be supporting evidence in the sense to be discussed below; i.e., *q* may be a presumptive factor for *p*, although the likelihood of *p* given *q* is *less* than the likelihood of *p* without *q* (the *a priori* likelihood of *p*). However, it can be shown that this can only happen in the case of statements (*p*) which are likely *a priori*, i.e., have *a priori* likelihood in excess of .5.

these conditions, can be introduced. This would be a measure —
here designated "dep" — having the characteristics embraced in
the following conditions:

Criteria of Adequacy

(DEP 1) dep (p, q) is a function of the quantities $Pr(p, q)$ and
$Pr(q)$

(DEP 2) $0 \leqslant$ dep $(p, q) \leqslant 1$, for any p, q

(DEP 3) If $Pr(q) = 1$, then dep (p, q) is $\leqslant$ or $\geqslant 0.5$ according
as $Pr(p, q)$ is $\leqslant$ or $\geqslant 0.5$

(DEP 4) If $Pr(p, q) = 0$, then dep $(p, q) = 0$

(DEP 5) If $Pr(p, q) = 1$, then dep $(p, q) = Pr(q)$

(DEP 6) If $Pr(q) = Pr(r)$ and $Pr(p, q) \geqslant Pr(p, r)$, then
dep $(p, q) \geqslant$ dep (p, r)

On the basis of these criteria of adequacy, a *dep*-measure can
be established as follows: (i) Let $x = Pr(p, q)$ and $y = Pr(q)$.
Then (DEP 1) is satisfied by defining dep (p, q) as a function
$F(x, y)$ of x and y. (ii) By (DEP 6) the function F is monotonic
in x. The *simplest* assumption, therefore, is that F is *linear*
in x, i.e., that $F(x, y) \equiv x \cdot G(y) + H(y)$. (iii) By (DEP 5),
$F(1, y) \equiv y$, and so $G(y) + H(y) \equiv y$ or $H(y) \equiv y - G(y)$.
Therefore in general, $F(x, y) \equiv x \cdot G(y) + y - G(y) \equiv y +
(x - 1)G(y)$. (iv) By (DEP 4), $0 \equiv F(0, y) \equiv y - G(y)$.
Therefore $G(y) \equiv y$. Thus by (iii) we have $H(y) \equiv 0$. Therefore
$F(x, y) \equiv x \cdot y$, i.e., dep $(p, q) = Pr(p, q) \cdot Pr(q)$. The reader can
readily check that the measure dep $(p, q) = Pr(p, q) \cdot Pr(q)$
satisfies all of the conditions of adequacy.

Abstracting from the likelihood of q itself, and thus resisting,
for the moment, the temptation to "simplify" the product
$Pr(p, q) \cdot Pr(q)$ to $Pr(p \& q)$, it can be seen that $Pr(p, q)$ measures
the *relative* degree of evidential presumption, relative to the
hypothesis q accepted as given. This shows that *the measure of
degree of confirmation as treated in the extensive literature on
confirmation is based upon the mode of evidential backing at
issue in the concept of evidential presumption.*

As mentioned earlier, two related concepts of evidence *supporting* evidence and *confirming* evidence are the focal points of the present study. I turn now to an examination of these concepts, and to the establishment of a measure of *degree of evidential support* analogous to the degree of evidential presumption which has been explicated.

By evidential presumption a statement is rendered *more likely than not*. Supporting evidence, on the other hand, renders a statement *more likely than before*, i.e., more likely *a posteriori* than it was *a priori*. Thus a (well-established) statement q may be said to be *supporting* evidence for the statement p if the conditional likelihood of p on q is greater than (or equal to) the absolute likelihood of p: $Pr(p, q) \geq Pr(p)$.

A numerical measure of *degree of evidential support* can be introduced. This measure — here designated "des" — should have the characteristics embraced in the following conditions:

Criteria of Adequacy

(DES 1) des (p, q) is a function of the quantities $Pr(p, q)$, $Pr(p)$ and $Pr(q)$

(DES 2) $-1 \leq$ des $(p, q) \leq 1$, for any p, q

(DES 3) des (p, q) is $\leq$ or ≥ 0 according as $Pr(p, q)$ is $\leq$ or $\geq Pr(p)$

(DES 4) des $(\sim p, q) = -$des (p, q)

(DES 5) If $Pr(p, q) = 1$, then des $(p, q) = Pr(q)$

(DES 6) If $Pr(q) = Pr(r)$ and $Pr(p, q) \leq Pr(p, r)$, then des $(p, q) \leq$ des (p, r)

(DES 7) If $Pr(p, q) \geq Pr(p)$ and $Pr(p, q) = Pr(p, r)$ and $Pr(q) \leq Pr(r)$, then des $(p, q) \leq$ des (p, r)

On the basis of these criteria of adequacy, a *des*-measure can be established as follows: (i) Let $x = Pr(p, q)$, $y = Pr(q)$ and $z = Pr(p)$. Then (DES 1) is satisfied by defining des (p, q) as a function $F(x, y, z)$ of x, y, and z. (ii) By (DES 7) we have that for fixed $x \geq z$, F is monotonic in y. The *simplest* assumption, therefore, is that for $z \leq x$, F is *linear* in y, i.e., $F(x, y, z) =$

$y \cdot G(x, z) + H(x, z)$. (iii) By (DES 3) we have that $F(x, y, z) \equiv$ 0 if $x = z$, i.e., if $Pr(p \& q) = y \cdot z$. Since $Pr(p \& q) \leqslant Pr(q) = y$, we have that if $y = 0$, $F = 0$. Therefore $H \equiv 0$, and $F(x, y, z) \equiv$ $y \cdot G(x, z)$. (iv) Under the assumption that $z \leqslant x$, so that (ii) and (iii) apply, we now use (DES 6) to remark that for fixed y, G is monotonic in x. Again, the *simplest* assumption is that G is linear in x, i.e., $G(x, z) \equiv x \cdot f(z) + g(z)$. (v) Thus by (DES 3) we have that if $z \leqslant x$, $G(z, z) \equiv z \cdot f(z) + g(z) \equiv 0$. Therefore $g(z) \equiv$ $-z \cdot f(z)$, and so we have that in general $G(x, z) \equiv f(z) \cdot (x - z)$. (vi) Still assuming that $z \leqslant x$, so that the previous steps apply, we have by now (DES 5) that $y \equiv y \cdot G(1, z) \equiv y \cdot f(z) \cdot (1 - z)$, so that $f(z) \equiv 1/(1 - z)$. Therefore if $z \leqslant x$, $F(x, y, z) \equiv (x - z)/$ $(1 - z) \cdot y$. (vii) It remains to deal with the case $z \geqslant x$. In this case we can use the above to compute

$$\text{des}(\sim p, q) = \frac{(1 - x) - (1 - z)}{1 - (1 - z)} \cdot y = \frac{z - x}{z} \cdot y.$$

Therefore, using (DES 4), we have that $\text{des}(p, q) =$ $[(x - z)/z] \cdot y$ in this case.

The stated conditions of adequacy thus lead to the measure

$$\text{des}(p, q) = \frac{Pr(p, q) - Pr(p)}{1 - Pr(p)} \cdot Pr(q)$$

or

$$\frac{Pr(p, q) - Pr(p)}{Pr(p)} \cdot Pr(q)$$

according as $Pr(p, q)$ is $\geqslant$ or $\leqslant Pr(p)$. The reader can readily check that this measure will satisfy all of the criteria of adequacy.[19]

Abstracting from the likelihood of q itself, it is seen that the ratio of $Pr(p, q) - Pr(p)$ to $Pr(\sim p)$ or $Pr(p)$, according to $Pr(p, q)$ is $\geqslant$ or $\leqslant Pr(p)$, measures the *relative* degree of evidential support, relative to the hypothesis q accepted as given. This quantity may be said to measure the *evidential relevance*

19. This measure was devised by the writer in the course of a collaborative investigation with Olaf Helmer, of evidential reasoning in the social sciences.

of q to p, for it is a measure of how much the credibility of p is decreased or increased when q is given.[20] The nature of this measure is indicated by the graphs in Figure II-1.

20. In recent years, considerable attention has been given by logicians to the development of measures of the *content* of a statement, and of the degree of *relevance* of one statement to another. Some major contributions to the discussion of content measures are (1) L. Wittgenstein, *Tractatus Logico-Philosophicus* (1922), (2) K. R. Popper, *Logik der Forschung* (1935), (3) C. G. Hempel and P. Oppenheim, "Studies in the Logic of Confirmation," *Philosophy of Science*, vol. 15 (1948), pp. 135–175, (4) R. Carnap, *Logical Foundations of Probability* (1950), and (5) J. G. Kemeny, "A Logical Measure Function," *The Journal of Symbolic Logic*, vol. 18 (1953), pp. 289–308. This investigation of measures of the content of *a statement*, while closely related and relevant to a discussion of relevance *among statements*, covers a separate portion of ground.

Major contributions to the discussion of relevance measures include (1) J. M. Keynes, *Treatise on Probability* (1921), (2) items number 2 and (3) number 4 above, (4) J. G. Kemeny and P. Oppenheim, "Degree of Factual Support," *Philosophy of Science*, vol. 19 (1952), pp. 307–324, (5) K. R. Popper, "Degree of Confirmation," *British Journal for the Philosophy of Science*, vol. 5 (1954), pp. 143–149. (See also a review of the last-named paper by Kemeny in *The Journal of Symbolic Logic*, vol. 20 [1955], pp. 304–305.)

In the 1950's lively discussion broke out in the *British Journal for the Philosophy of Science* regarding measures of confirmatory relevance among statements measures is of interest in revealing relationships and in illustrating the many (among other things). The discussants included Y. Bar-Hillel (vol. 6 [1955], pp. 155–157; vol. 7 [1956], pp. 245–248), K. R. Popper (vol. 6 [1955], pp. 157–163; vol. 7 [1956], pp. 244–245 and 249–256), and R. Carnap (vol. 7 [1956], pp. 243–244). This discussion, insofar as presently relevant, has dealt with the serviceability of alternative relevance measures for various uses in the theory of confirmation.

None of the relevance measures considered in this literature correspond in either intent or substance to the concept of evidential relevance as understood and explicated here. None the less, a comparison among the principal relevance-different forms of the relevance idea. The starting point of such a comparison is provided by Kemeny in the review cited above, and the upshot is essentially as follows: (A) Carnap is concerned to analyze the measure inherent in the question, "How sure are we of p if we are given q as evidence?" (B) Popper and Kemeny–Oppenheim deal with the question, "How much surer are we of p given q then without q?" (C) The present measure of evidential relevance deals with the question, "How much is our confidence in the truth of p increased or decreased if q is given?" Various alternative approaches to the measurement of evidential relevance are surveyed in an illuminating way in Henry E. Kyburg, Jr., article on "Recent Work in Inductive Logic," *American Philosophical Quarterly*, vol. 1 (1964), pp. 249–287 (see pp. 255–257).

The function $R = \begin{cases} \dfrac{x-z}{1-z} \\[2mm] \dfrac{x-z}{x} \end{cases}$ according as $x \begin{cases} \geqslant z \\ \leqslant z \end{cases}$

$$x = L(p, q) \qquad\qquad z = L(p)$$

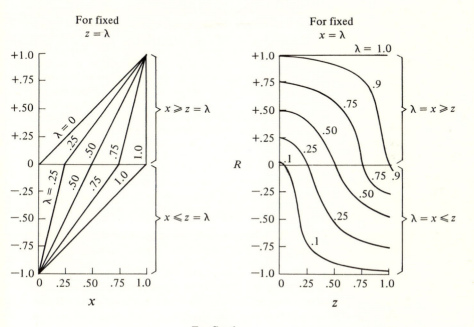

For fixed $z = \lambda$

For fixed $x = \lambda$

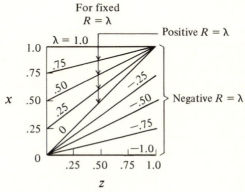

For fixed $R = \lambda$

Figure II-1

We can now finally introduce our specification of the concept of confirming evidence. The statement p is *confirming evidence* for q if (1) p is a presumptive factor for q, and (2) p is supporting evidence for q. Thus confirming evidence must at once render its hypothesis more likely than before *and* more likely than not. This two-pronged concept of confirming evidence perhaps most closely approximates to the idea represented by the common usage of "evidence."

Having analyzed the nature of the evidence concept and distinguished various significantly different modes of evidence, let us now turn to the formal logic of the evidential relation, conceived as a logical relationship between statements. Three classes of rules will be of primary interest: *rules for confirming evidence* which are valid for both evidential presumption and for supporting evidence, and *special rules* which are valid for only one of these subordinate evidential concepts. In addition, there are certain rules which may appear plausible at first sight but do not hold for any of these senses of evidence. A number of these *illicit rules* will also be listed.

A formal theory of evidence is an addition to the logical theory of necessary inference, and represents an extension of such a theory to encompass certain less rigid and demanding modes of inference. Here, then, a system of deductive logic is presupposed, and additional rules for the evidential relationship—to be denoted by the symbol "E"—will be conjoined to this system.

The acceptability of a proposed rule of evidence as applicable to the two relationships of evidential presumption and of supporting evidence will be *tested* by use of the explication of these concepts in terms of the likelihood measure Pr. On our approach the *unqualified* acceptability of a rule to confirming evidence will be construed to amount to its applicability to *both* of these. In this testing process, the interpretation of "E" is such that

(i) For *evidential presumption* pEq amounts to $Pr(q,p) \geqslant Pr(\sim q,p)$, i.e., $Pr(p \mathbin{\&} q) \geqslant Pr(p \mathbin{\&} \sim q)$

(ii) For *supporting evidence pEq* amounts to $Pr(q,p) \geqslant Pr(q)$, i.e., $Pr(p \& q) \geqslant Pr(p) \cdot Pr(q)$.[21]

The validity of proposed rules of evidence will be tested by these interpretations, by checking if the rule in question is valid for any possible assignment of *Pr*-values to the statements involved.[22]

1. *Rules for Confirming Evidence.* A rule for confirming evidence is a thesis that is valid for each legitimate interpretation of the evidence concept. Some specific examples of rules of this type are as follows. (We use the standard symbol ⊢ for the relationship of logical consequence.)

(RCE 1) Logically equivalent statements are intersubstitutable in evidence contexts. (This is properly a *rule of inference*.)

(RCE 2) A statement which entails another is evidence for it:

$$(p \rightarrow q) \vdash (pEq)$$

(RCE 3) A consistent statement is never evidence for a statement with which it is incompatible:

$$[\sim (p \rightarrow \sim p) \& (p \rightarrow \sim q)] \vdash \sim (pEq)$$

Each of these rules can be shown to apply to both evidential presumption and supporting evidence. To illustrate the procedure

21. Actually, some suitable inequality of the form $Pr(q) \geqslant \xi$ is also supposed in each of these cases, but no use will be made of this fact. This additional condition may, of course, be of great significance in applications to uses of evidence (e.g., in law) for which it is important to impose the requirement that confirming evidence be significantly greater than 0.5. However, the condition will not be needed here, because it has no bearing on the logical form or structure of rules of evidence, this matter alone being presently under discussion.

22. In the formulation of rules of evidence, no attempt is made to attain *completeness*, i.e., to ensure that all rules meeting the respective criterion of acceptability are logical consequences of the given list of basic rules. The formulation of complete axiom systems for the three evidence-relations considered below (the relationships *evidential presumption* and of *confirming* and of *supporting* evidence) is left as an open problem.

for testing such rules, the test of (RCE 2) is exhibited as a sample:

TEST 1: APPLICABILITY TO EVIDENTIAL PRESUMP-TION. Let

$$Pr(p \& q) = x_1 \qquad Pr(\sim p \& q) = x_3$$

$$Pr(p \& \sim q) = x_2 \quad Pr(\sim p \& \sim q) = x_4$$

Now (1) $p \rightarrow q$ iff (if and only if) $x_2 = 0$. And (2) pEq in the sense of evidential presumption iff $Pr(p \& q) \geq (p \& \sim q)$, i.e., iff $x_1 \geq x_2$. Since x_1 is necessarily ≥ 0, (1) implies (2), and (RCE 2) is thus valid for evidential presumption.

TEST 2: APPLICABILITY TO SUPPORTING EVIDENCE. Let the x_i be defined as above, so that again (1) $p \rightarrow q$ iff $x_2 = 0$. Now (2) pEq in the sense of supporting evidence iff $Pr(p \& q) \geq Pr(p) \cdot Pr(q)$, i.e., iff $x_1 \geq (x_1 + x_2) \cdot (x_1 + x_3)$. Since $1 \geq x_1 + x_3$, and thus $x_1 \geq x_1 \cdot (x_1 + x_3)$, we have that (2) follows from (1). Thus (RCE 2) is also valid for supporting evidence.

The two other rules can be tested in a wholly analogous manner.

From these three basic rules, certain others follow as logical consequences. Several of these are listed. No proofs are given, the proof being elementary in all cases.

(RCE 4)   Any statement is evidence for itself:

$$pEp$$

(RCE 5)   A conjunction is evidence for its members:

$$(p \& q)Ep$$

(RCE 6)   An inconsistent statement is evidence for any state-ment:

$$(p \rightarrow \sim p) \vdash (pEq)$$

(RCE 7)   Any statement is evidence for a necessary statement:

$$(\sim p \rightarrow p) \vdash (qEp)$$

(RCE 8)   No consistent statement is evidence for an inconsistent statement:

$$[\sim (p \rightarrow \sim p) \,\&\, (q \rightarrow \sim q)] \vdash \sim (pEq)$$

Each of these propositions is also a general rule of evidence, valid for each of the various constructions of the evidence concept as already analyzed.

2. *Special Rules for Evidential Presumption.* A *special rule of evidence* is one that is valid for one interpretation of the subordinate evidential concepts, but not the other. We first consider some special rules for evidential presumption:

(REP 1)   If a statement is evidence for its own negation, it is inconsistent:

$$(pE \sim p) \vdash (p \rightarrow \sim p)$$

(REP 2)   If a statement is evidence for each of two statements, it is evidence for their disjunction:

$$[(pEq) \,\&\, (pEr)] \vdash [pE(qVr)]^{23}$$

(REP 3)   If a statement is evidence for another, it is evidence for any statement entailed by this statement:

$$[(pEq) \,\&\, (q \rightarrow r)] \vdash (pEr)$$

Each of these rules can be shown to hold for evidential presumption but to fail to apply to supporting evidence. A sample verification of this is here provided for (REP 2).

23. It is worthwhile to contrast this rule with two cognates:

$$[(pEq) \,\&\, (pEr)] \vdash [pE(q \,\&\, r)]$$

$$(pvq)Ep$$

The former *rule of conjunction* perhaps rather surprisingly fails for both evidential presumption and supporting evidence. The second rule obtains with respect to supporting evidence but fails for evidential presumption.

TEST 1: APPLICABILITY TO EVIDENTIAL PRESUMP-
TION. Let

$$Pr(p \mathrel{\&} q \mathrel{\&} r) = x_1 \qquad Pr(\sim p \mathrel{\&} q \mathrel{\&} r) = x_5$$
$$Pr(p \mathrel{\&} q \mathrel{\&} \sim r) = x_2 \qquad Pr(\sim p \mathrel{\&} q \mathrel{\&} \sim r) = x_6$$
$$Pr(p \mathrel{\&} \sim q \mathrel{\&} r) = x_3 \qquad Pr(\sim p \mathrel{\&} \sim q \mathrel{\&} r) = x_7$$
$$Pr(p \mathrel{\&} \sim q \mathrel{\&} \sim r) = x_4 \quad Pr(\sim p \mathrel{\&} \sim q \mathrel{\&} \sim r) = x_8$$

Now, taking E in the sense of evidential presumption we have:

$$pEq \text{ iff } x_1 + x_2 \geqslant x_3 + x_4 \tag{1}$$
$$pEr \text{ iff } x_1 + x_3 \geqslant x_2 + x_4, \text{ and finally} \tag{2}$$
$$pE(q \lor r) \text{ iff } x_1 + x_2 + x_3 \geqslant x_4 \tag{3}$$

But (3) is readily seen to follow from (1) and (2) together. Thus
(REP 2) is valid for evidential presumption.

TEST 2: INAPPLICABILITY TO SUPPORTING EVI-
DENCE. Again, let the x_i be defined as above. Taking E in
the sense of supporting evidence we have:

$$pEq \text{ iff } x_1 + x_2 \geqslant (x_1 + x_2 + x_3 + x_4)(x_1 + x_2 + x_5 + x_6) \tag{1}$$
$$\text{iff } x_1 + x_3 \geqslant (x_1 + x_2 + x_3 + x_4)(x_1 + x_3 + x_5 + x_7) \tag{2}$$
and
$$pE(qVr) \text{ iff } x_1 + x_2 + x_3 \geqslant (x_1 + x_2 + x_3 + x_4)(x_1 + x_2 + x_3 + x_5$$
$$+ x_6 + x_7) \tag{3}$$

That (3) does not in general follow from (1) and (2) may be seen
from the following assignment of values to x_1 through x_8, respectively: 0.1, 0, 0, 0.2, 0.1, 0.1, 0.1, 0.4. Thus (REP 2) is not valid
for supporting evidence.

The remaining rules can be checked in an analogous manner.

Among the logical consequences of these rules [added to the
RCE's], two deserve listing here as additional REP's:

(REP 4) If a statement is evidence for one of two statements,
it is evidence for their disjunction:

$$(pEq) \vdash [pE(q \lor r)]$$

(REP 5) If a statement is evidence for a conjunction, it is evidence for each term:

$$[pE(q \ \& \ r)] \vdash (pEq)$$

These two rules also are applicable solely to evidential presumption.

3. *Special Rules for Supporting Evidence*. Analogously with the special rules for evidential presumption, there are also rules applicable only to supporting evidence. An example is

(RSE 1) If one statement is evidence for another, this second statement is evidence for the first:

$$(pEq) \vdash (qEp)$$

The reader can readily check that this statement is valid for supporting evidence, but does not hold for evidential presumption.[24]

24. The question of the mutual or reciprocal evidence of propositions for each other has some interest. (RSE 1) settles this matter as regards supporting evidence. With respect to evidential presumption, I offer the following calculation:

$$\text{dep}(p,q) = Pr(p,q), \text{ and dep}(q,p) = Pr(q,p). \tag{1}$$

By Bayes's Theorem,

$$Pr(p,q) = \frac{Pr(q,p) \cdot Pr(p)}{Pr(q,p) \cdot Pr(p) + Pr(q, \sim p) \cdot Pr(\sim p)}. \tag{2}$$

By (2) and the rules of "Pr,"

$$Pr(p,q) = \frac{Pr(q,p) \cdot Pr(p)}{Pr(q)}. \tag{3}$$

By (1) and (3), we have

$$\text{dep}(p,q) = \text{dep}(q,p) \cdot \frac{Pr(p)}{Pr(q)}. \tag{4}$$

We are not warranted in saying more regarding the relationship of dep(p,q) and dep(q,p) than is contained in (4), viz., that evidential presumption is a fully reciprocal relationship only among equally likely statements.

Among the logical consequences of this rule (added to the RCE's), the following two may be listed as additional RSE's:

(RSE 2) If one statement entails another, this second statement is evidence for the first:

$$(p \rightarrow q) \vdash (qEp)$$

(RSE 3) One term of a conjunction is evidence for the whole:

$$pE(p \& q)$$

These two rules hold for supporting evidence but do not apply to evidential presumption.

Using as building blocks the rules of evidence which have been discussed, three logical systems can be constructed:

 i. A Theory of Confirming Evidence based only upon the rules of evidence applicable to each of the subordinate evidence concepts.

 ii. A Theory of Evidential Presumption obtained by supplementing the rules for confirming evidence by the special rules for evidential presumption, and

 iii. A Theory of Supporting Evidence obtained by supplementing the rules for confirming evidence by the special rules for supporting evidence.

No further elaboration of these theories is undertaken here. But one closing remark regarding these systems of rules of evidence is in order. From the logical standpoint of systematic richness and power, it would seem that the stricter criterion of acceptability for rules of confirming evidence is a curtailing factor in limiting the range of acceptable assertions. This system is, however, the most interesting from the standpoint of its application or interpretation, because *confirming evidence* most closely fits the broad outlines of the intuitive conception of evidence, and because only those rules which hold for all of the subordinate evidence concepts be taken to apply to the concept of evidence *per se*.

4. *Illicit Rules of Evidence.* Going beyond the special rules of evidence, we meet with certain other statements involving the evidence relation which may appear on first view to merit inclusion in a logical theory of evidence, but which can be seen upon closer scrutiny to hold for none of the evidence concepts which we have here been considering. The following are some examples of such illicit rules of evidence:

(IRE 1) If p is evidence for q, and r entails p, then r is evidence for q:
$$[(pEq) \ \& \ (r \rightarrow p)] \vdash (rEq)$$

(IRE 2) If p is evidence for q, then not-q is evidence for not-p:
$$(pEq) \vdash (\sim qE \sim p)$$

(IRE 3) If p is evidence for q, and q is evidence for r, then p is evidence for r:
$$[(pEq) \ \& \ (qEr)] \vdash (pEr)$$

(IRE 4) If each of two statements is evidence for a given statement, so is their disjunction:
$$[(pEq) \ \& \ (rEq)] \vdash [(p \lor r)Eq]$$

(IRE 5) If a consistent statement is evidence for each of two statements, these are mutually compatible:
$$[\sim (p \rightarrow \sim p) \ \& \ (pEq) \ \& \ (pEr)] \vdash \ \sim (q \rightarrow \sim r)$$

(IRE 6) No true statement is evidence for a false statement. If a statement is true, and is evidence for another, then this second statement is also true:
$$[p \ \& \ (pEq)] \vdash q$$

(IRE 7) If a statement is evidence for another, and also for the negation of this statement, the original statement is inconsistent:
$$[(pEq) \ \& \ (pE \sim q)] \vdash (p \rightarrow \sim p)$$

The refutation of each of these can be carried through both for evidential presumption and for supporting evidence by the construction of counterexamples (i.e., falsifying assignments of *Pr*-values). Thus all three theories of evidence must exclude these rules. With regard to (IRE 1), it should be remarked that if *r* is highly confirmed — i.e., $Pr(\sim r)$ is essentially 0 — this rule does hold as a rule for confirming evidence. Analogously, similar special conditions can be found which render some of the other "illicit rules" acceptable. Herein lies the seemingly paradoxical character of the illicit status of *some* of these rules. They are valid, not in general, but only under special conditions which in practice are tacitly assumed to be satisfied.

* * *

It is proper at this point to mention two precursors of the present approach to the development of a logical theory of rules of evidence. In the course of a study of the formal logic of confirmation, C. G. Hempel has presented a list of "criteria of adequacy" for a theory of confirmation.[25] Hempel's criteria can in fact be properly and appropriately regarded as a set of rules for the evidential relation. In Carnap's *Logical Foundations of Probability*, Hempel's rules are subjected to a critical examination, and a modified set of rules presented.[26]

Hempel's "criteria" include several of the rules of confirming evidence listed previously, but in addition include also (REP 2), (REP 3), and (IRE 5). Thus, dismissing Hempel's acceptance of (IRE 5) as an inadvertent error, it becomes plausible to suppose that Hempel is in fact concerned to formulate a set of rules applicable to the idea of *evidential presumption* as here understood.

It is primarily on the score of the inclusion of (REP 2), (REP 3), and (IRE 5) that Hempel's list of criteria was criticized in the

---

25. "A Purely Syntactical Definition of Confirmation," *The Journal of Symbolic Logic*, vol. 8 (1943), pp. 122–143. "Studies in the Logic of Confirmation," *Mind*, vol. 45 (1945), pp. 1–26, 97–121.

26. Pp. 468–482 give a critique of Hempel's treatment of evidence.

examination made of it in Carnap's book. Carnap explicitly rejects these three rules, otherwise concurring — except for one or two minor points — in Hempel's discussion. Carnap's evaluation of the acceptability of rules of evidence is not informal and intuitive (like Hempel's), but is based upon an arithmetized criterion of acceptability analogous to those used in the present paper. However, Carnap's test for the acceptability of rules of evidence (accord with every regular c-function, as this is defined in his book) amounts to this: that the rule is valid if pEq is interpreted as $C(q, p) > C(q)$, where C is a certain type of Pr-measure. Thus the evidence-concept upon which Carnap's discussion seems to be based is closely akin to the concept of *supporting evidence* as understood here.[27] This difference in approach to the construction of the concept of evidence seems best to account for the points of disagreement at issue between Hempel and Carnap: these authors are apparently concerned to give rules for *different conceptions of evidence*, and so it is not surprising that they should differ regarding the appropriateness of various rules.

In any event, this brief comparison of the rules of evidence considered here and the conditions for confirming evidence as discussed by Hempel and Carnap shows the kinship of the present theory of evidence with their work.

* * *

The central and fundamental fact of the theory of evidence is that one statement may constitute evidence for another which goes beyond it in content. This feature fundamentally differentiates the concept of *evidence* from that of deductive *entailment:* the comparative weakness of the former embodying its very reason for being. A true statement may legitimately provide evidence for a falsehood; a statement may constitute evidence for each of several incompatible statements (cf. (IRE 5) and (IRE 6), and the motto from Bishop Butler).

27. In particular, all RCE's and all RSE's would be acceptable to Carnap.

Thus in domains in which one operates with evidence proper, leaving behind the secure ground of proof, the possibility of error must be accepted. Indeed in such fields a valid distinction may be drawn between *error* (drawing a conclusion, which, though false, actually derives from a soundly conducted inquiry) and *mistake* (falsehood owing to a fallacy in the inquiry).

This characteristic of evidence points also to a methodological contrast between deductive theoretical systems, where the relevant species of evidence furnishes a complete basis for the propositions it supports, and theoretical systems where slippage between evidence and conclusion may legitimately occur. Deductive systems can afford to be completely forward-looking, in that it is never appropriate to employ a proposition newly arrived at to give additional bolstering and support to one attained at some previous juncture. Other evidential systems, however, cannot always afford to forego retrospective arguments. The looser the evidence concept appropriate to an area of inquiry, the more will the system of reasoning take the form of a cluster of interlocking propositions lending mutual strength and support for one another. Such theoretical systems can range from those based upon formal canons of evidential arguments, as with legal evidence, to systems admitting the most tentative and provisional modes of argument. However, no system of this type presents the aspect of a collection of chains of deduction. Rather, they are akin to crossword puzzles, each piece bolstering and interlocking with every other.

In such a sphere in which the cutting edge of the evidence concept is less keen, legitimate disagreement can arise regarding the truth of statements in the face of agreement regarding the evidential basis for these statements. Despite agreement on "the facts" different constructions or interpretations can be placed upon them — particularly if the scope of discussion is speculative, i.e., on a level at several removes in generality and abstraction from "the facts." In a context of this sort, the common requirement that a statement agree with "the facts" ceases to be an effectual criterion of validity, because a significant area of assertion lies beyond the discriminatory power of this criterion. Our common conceptions regarding the acceptability of evidential reasonings, insofar as they derive from nurture in the areas of demonstrative reasonings in mathematics and classical physics are not an unfailing guide. They should not be carried over uncritically into logically less tidy areas where — by virtue of the applicability of a fundamentally different type of evidence — they have no place.[28]

28. Cf. Aristotle, *Ethics*, I, 3.

part III

PHILOSOPHICAL ISSUES OF SCIENTIFIC EXPLANATION

1. What Is a Universal Law? The Nature of Lawfulness

Scientific explanations are subsumption arguments: they place special cases within a framework of regularity represented by laws. Our concept of explanation — causal explanation pre-eminently included — is such as to require that the generalizations used for explanatory purposes must be *lawful*. An account of the nature of lawfulness is thus a central task of any adequate theory of scientific explanation.

Now it is quite clear that not just any universal empirical generalization will qualify as a law in this scientific context of discussion, no matter how well established it may be. It is critically important to distinguish here between *accidental* generalizations on the one hand and *lawful* generalizations on the other. "All coins in my pocket weigh less than one ounce" and "All American presidents are natives of the continental United States" are examples of accidental generalizations. By contrast, generalizations like "All elm trees are deciduous," "All (pure) water freezes at 32°F," and "All *Y* chromosomes self-duplicate under stimulation" are lawful. An accidental generalization claims merely that something *is* so—perhaps even that it is always so, whereas a lawful generalization claims that something *must* (in some appropriate sense) be so.

Thus consider the following two answers to the explanatory question, "Why did that tree shed its leaves last fall?"

(1) Because it is an elm, and *all elms are deciduous.*
(2) Because it is a tree in Smith's yard, and *all trees in Smith's yard are deciduous.*

The drastic difference in the satisfactoriness of these two "explanations" is due exactly to the fact that the generalization deployed in the first is lawful, whereas that in the second is not. Laws are akin to, yet different from both rules and descriptions. Like rules, laws state how things "must be"; yet unlike most familiar sorts of rules, laws admit no exceptions, but are always obeyed. Like descriptions, laws state how things are; yet unlike standard descriptions laws go beyond describing how things in fact are to make claims about how they must be. Thus laws have both a descriptive and a rulish aspect that prevents their being grouped smoothly into either category.

But just what is this factor of lawfulness that is present with generalizations and absent with others? The best way to answer this question of what lawfulness *is*, is by inquiring into what it *does*. Lawfulness manifests itself in two related ways: *nomic necessity* and *hypothetical force*. Nomic necessity introduces the element of *must*, of inevitability. In asserting it *as a law* that "All *A*'s are *B*'s" ("All timber wolves are carnivorous") we claim that the world being as it is it is necessary that an *A* must be a *B* (i.e., that a timber wolf will under appropriate circumstances unfailingly develop as a meat-eating animal).

This nomic necessity manifests itself most strikingly in the context of hypothetical suppositions, especially counterfactual hypotheses. In accepting "All *A*'s are *B*'s" ("All spiders are eight-legged") as a law, we have to be prepared to accept the conditional

"If x were an A, then x would be a B." (If this beetle were a spider — which it isn't —, then it would have eight legs.) It is pre-eminently this element of hypothetical force that distinguishes a genuinely lawful generalization from an accidental generalization like "All coins in my pocket weigh less than one quarter ounce." For we would not be prepared to accept the conditional "If this Venetian florin were a coin in my pocket, then it would weigh less than one quarter ounce."[1]

The fact is that the statement

 (1) All X's are Y's

makes a stronger claim when put forward as a law than when put forward as a "mere" generalization. For if (1) is construed as a law it asserts "All X's *have to be* Y's," that is, we obtain the stronger *nomological* generalization:

 (1a) All X's are Y's *and further* if z (which isn't an X) were an X, then z would be a Y.[2]

When a generalization is taken as lawful it obtains added force; it gains an added assertive increment, even though this nomic necessity will express itself primarily in applications of a counter-factual kind. For it is clearly in hypothetical and counterfactual contexts that nomic necessity manifests itself most strikingly.

Consider the counterfactual supposition: *Assume that this wire*

---

  1. It is clear that we mean this to be construed as "if it were somehow *added to* the coins in my pocket" and not as "if it were to be somehow *identical with* one of the coins in my pocket."

  2. Roderick M. Chisholm has put this point with admirable precision: Both law statements and non-law statements may be expressed in the general form, "For every x, if x is an S, x is a P." Law statements unlike non-law statements, seem "however" to warrant inference to statements of the form, "If a, which is not S, were S, a would be P" and "For every x, if x were S, x would be P." R. M. Chisholm, "Law Statements and Counterfactual Inference," *Analysis*, vol. 15 (1955), p. 97.

(which is actually made of copper) is made of rubber. This supposition occurs in the following context:

ITEMS OF KNOWLEDGE

Facts: (1) This wire is made of copper.
         (2) This wire is not made of rubber.
         (3) This wire conducts electricity.
         (4) Copper conducts electricity.
Laws: (5) Rubber does not conduct electricity.

HYPOTHESIS

Not (2), i.e., This wire is made of rubber.

To restore consistency in our knowledge in the face of this hypothesis we must obviously give up (1) and (2). But this is not sufficient. One of (5) or (3) must also go, so that *prima facie* we could adopt either of the conditionals:

(A) If this (copper) wire were made of rubber then it would not conduct electricity (because rubber does not conduct electricity).

(B) If this (copper) wire were made of rubber then rubber would conduct electricity (because this wire conducts electricity).

That is, we get a choice between retaining (5) with alternative (A) and retaining (3) with alternative (B). It is precisely because classing a statement as a "law" represents an epistemic commitment to retain it in the face of counterfactual hypotheses that the conditional (A), viz., "If this wire were made of rubber then it would not conduct electricity" strikes us as natural vis-à-vis (B).[3]

One effective way to motivate the distinction between a "law" and a "mere generalization" is to consider the effect of the logical process of contraposition. As we have seen, the statement

(1) All *X*'s are *Y*'s

makes a stronger claim when put forward as a law than when put

---

3. The considerations at issue here are treated in more detail in my book on *Hypothetical Reasoning* (Amsterdam, 1965).

forward as a "mere" generalization. For if (1) is construed as a law it asserts "All X's *have to be* Y's," so that we obtain the stronger *nomological* generalization:

(1a) All X's are Y's *and further* if z (which isn't an X) were an X, then z would be a Y.

When a generalization of the type (1) is taken in this nomological way, contraposition clearly fails. For

(2) All non-Y's are non-X's

when construed as stating a law, will assert:

(2a) All non-Y's are non-X's *and further* if z (which isn't a non-Y) were a non-Y then z would be a non-X.

Although the generalizations (1) and (2) are equivalent, this is not the case with their nomological counterparts (1a) and (2a). (These statements are nonequivalent because, *inter alia*, (1a) affirms that the z at issue in it is to be a non-X, whereas (2a) affirms that the z at issue in it is to be a Y [and neither statement justifies relating non-X's and Y's].) Thus _seemingly equivalent_ _generalizations can formulate different, nonequivalent laws_. When a generalization is taken as nomological, that is as stating a law, it obtains an assertive increment (albeit one of a strictly counterfactual sort) of such a kind that contraposed generalizations will no longer represent the same law.

It is worth noting, incidentally, that these considerations help also to shed some light upon the inductive procedures by which laws are verified. For the purposes of inductive confirmation (1a) comes to be reconstrued in strictly factualistic, albeit epistemologized terms. It is reoriented from the realm of *counterfact* to that of *ignorance* (absence of information), being rendered epistemological rather than counterfactual by means of the indicated italicized insertions:

(1b) All X's are Y's and further if z (which isn't *known to be* an X) were *to turn out* to be an X, then z would *turn out to* be a Y.

Thus in verifying (1) the *prima facie* procedure is to hunt X's and check that they are Y's. On the other hand (2) gives rise to:

(2b) All non-Y's are non-X's and if z (which isn't known to be a non-Y) were to turn out to be a non-Y, then z would turn out to be a non-X.

And so in verifying (2) the *prima facie* procedure is to hunt for non-Y's and check that they are non-X's. And these procedures, *qua* procedures, are very different, *despite the fact that a counterexample provided by one of them must also turn out to be a counterexample for the other* (viz., an X that is a non-Y). For although these generalizations are *refuted* (counterinstanced) by the same negative findings, they differ — as we saw in the preceding section — in the amount of the inductive *support* (confirmation) which they derive from positive, nonrefuting instances of different kinds, and thus differ with respect to the operational procedures of verification.

It is built into our very concept of a law of nature that such a law must, if it be of the universal type, correspond to a universal generalization that is claimed to possess nomic necessity and is denied to be of a possible merely accidental status. If the generalizations were claimed to hold *in fact* for all places and times, even this would not of itself suffice for lawfulness: it would still not be a law if its operative effectiveness were not also extended into the hypothetical sphere. The conception of a universal law operative in our concept of causal explanation is thus very complex and demanding. A lawful generalization goes beyond the claims of a merely factual generalization as such; it involves claims not only about the realm of observed fact, but about that of hypothetical counterfact as well. And just these far-reaching claims are indispensable to the acceptance of a generalization as lawful and is a formative constituent of our standard concept of a universal law of nature.

That laws demand nomic necessity is a point regarding which there is a substantial consensus in the history of philosophy. Aristotle insists on the point in the *Posterior Analytics*.[4] It is a basic theme in Kant's *Critique of Pure Reason*.[5] And it continues operative in current philosophy. Such writers as C. J. Ducasse and A. Pap, for example, hold that natural laws involve a necessity that is not logical but yet transcends merely *de facto* regularity.[6] Both Nelson Goodman and Roderick M. Chisholm have proposed hypothetical force as a prime criterion of the nomic necessity requisite for lawfulness.[7] And nowadays it is a matter of widespread agreement that some characteristic mode of nomic necessity is involved in lawfulness, although writers differ as to just how the factor of nomic necessity is to be explicated. The writer of the relevant article in the most recent philosophical encyclopedia puts the matter accurately by saying that the current point of dispute "is not about the propriety of using such terms as 'nomic necessity,' rather it is about the interpretation of these terms or the justification of their use."[8]

Conceding hypothetical force as an ingredient in all *universal* laws of nature, one recent writer denies its applicability to lawful-

---

4. See especially sections 1–6 of Book I.

5. See especially the sections Introduction and Transcendental Analytic.

6. See Curt J. Ducasse, "Explanation, Mechanism, and Teleology," *The Journal of Philosophy*, vol. 23 (1926), pp. 150–155. Reprinted in H. Feigl and W. Sellars (eds.), *Readings in Philosophical Analysis* (New York, 1949). Arthur Pap, *An Introduction to the Philosophy of Science* (New York, 1962), see Chapter 16.

7. R. M. Chisholm, "The Contrary-to-Fact Conditional," *Mind*, vol. 55 (1946), pp. 289–307, reprinted in H. Feigl and W. Sellars (eds.), *Readings in Philosophical Analysis* (New York, 1949). Nelson Goodman, "The Problems of Counterfactual Conditionals," *The Journal of Philosophy*, vol. 44 (1947), pp. 113–128, reprinted in L. Linsky (ed.), *Semantics and the Philosophy of Language* (Urbana, Ill., 1952), and in N. Goodman, *Fact, Fiction, and Forecast* (Cambridge, Mass., 1955).

8. R. S. Walters, "Laws of Science and Lawlike Statements" in the *Encyclopedia of Philosophy* ed. by P. Edwards, vol. 4 (New York, 1967), pp. 410–414 (see pp. 411–412). This article offers a very clear and compact survey of the key issues regarding laws.

ness in general on the ground that it is lacking in statistical laws.[9]
Now if I use that statistical law "The half-life of californium 247
is 2.4 hours" as basis for explaining why a particular atom of
californium lasted $2\frac{1}{2}$ hours, then I cannot go on to say things like
"If this atom of uranium 235 had been californium 247, then it
would have lasted only $2\frac{1}{2}$ hours." This sort of specific and
nonprobabilistic application of a statistical law is indeed impossible. But we surely can and would be prepared to say things like
"If this atom of uranium 235 had been californium 247, then it
would have had a half-life of 2.4 hours, and so the probability
exceeds .85 that it would have lasted 4 ± 1 hours." Statistical laws
too can (and indeed *must*, if lawful) be capable of counterfactual
applications, only in their case such applications will, naturally
enough, take a probabilistic form.

Some recent writers have advocated a "regularity theory" of
laws according to which lawfulness is to be construed as unrestricted factual generality pure and simple, so that no trans-
inductive imputation of nomic necessity is called for. As R. S.
Walters puts the matter, the key

> objection to the regularity theory is that it cannot account for
> possible instances. If this charge were indeed well founded, it
> would be difficult to see how one could avoid the view that natural
> laws assert some kind of necessity such that they apply in all
> possible worlds. However, it is not established that a defender of
> the regularity view cannot give a plausible account of the applica-
> tion of laws to possible instances. He would argue that statements
> about possible instances stand in the same kind of logical relation
> to a law as do statements about actual unobserved instances. To
> the extent that a law enables prediction about unobserved in-
> stances, it enables justifiable claims about unrealized possibilities.[10]

9. Mario Bunge, *The Myth of Simplicity* (Englewood Cliffs, N.J., 1963),
p. 174. At least one influential adherent of the regularity theory was, however,
prepared to brush aside all reference to the possible, saying: "Physics wants to
establish regularities; it does not look for what is possible." [L. Wittgenstein in
his middle period as quoted by H. Spiegelberg in the *American Philosophical
Quarterly*, vol. 5 (1968), p. 256].

10. R. S. Walters, *op. cit.*, pp. 413–414.

But is is quite clear that this line of defense will not serve at all. It suffers from the critical defect of begging the question by treating the unobserved and the unactualized cases in exactly the same way, as different instances of the same thing. But it is quite clear on the basis of considerations we have already canvassed that this step is indefensible because the unobserved and the unreal are in a totally different position in the context of inductive considerations, since the realm of the (heretofore) unobserved is open to observational exploration whereas the domain of the hypothetically unreal lies *ex hypothesi* beyond our reach.

Of course one could try to argue that the consideration of hypothetical cases is improper or illicit (illegitimate, "beside the point," or whatever) — that reality alone concerns us and that the unreal lies wholly outside the sphere of legitimate consideration. This does indeed abrogate the difficult nomic aspect of lawfulness. But it also writes off the prospect of hypothetical reasoning in the sciences and abolishes the concept of explanation as it has in fact developed in the context of the Western tradition of scientific methodology.[11]

2. Lawfulness as Imputation

On what evidential basis does an empirical generalization acquire the nomic necessity and hypothetical force it requires to qualify as a law? However substantial this evidential basis may be, no matter how massively the observational evidence may be amassed and how elaborately the case developed, it is clear upon reflection that this evidential basis must always be grossly insufficient to the claim actually made when we class a generaliza-

11. This position is in fact taken by latter day idealists of the type of F. H. Bradley and Brand Blanshard who hold in effect the nomic necessity and logical necessity are indistinguishably one and the same, so that counterfactual hypotheses cannot be posed at all in any meaningful or coherent way.

tion as a law.[12] This becomes evident in part for the familiar reasons that while all such evidence relates to the past (and possibly the present), scientific laws invariably also underwrite claims about the future. It is, moreover, also clear from considering the conceptual nature of lawfulness, bearing in mind that observation and evidence always relate to what happens in fact, whereas laws invariably also underwrite claims of a hypothetical or counterfactual kind.

Let us consider this root insufficiency of the evidential basis for a law somewhat more closely. It is obvious that this basis will be *deductively insufficient* because the evidence inevitably relates to a limited group of cases while the applicability of the law is unrestricted. Moreover the evidential basis will also be *inductively insufficient.* For inductive procedures are designed to warrant the step from observed to unobserved cases, whereas a law — whose very lawfulness arrogates to it nomological necessity and counterfactual force — not only takes this inductive step from observed to unobserved cases, but also takes the added step from actual to hypothetical cases. The inductive justification of hypothetical force would have to take the form "has always been applicable to counterfactual cases." And the premiss for such an induction will obviously always be unavailable. The evidential foundation for generalization is thus afflicted by a double insufficiency, not only in the *deductive* mode, but also *inductively* (at any rate as long as induction is construed along usual and standard lines).[13]

12. For just this reason, major philosophers from Aristotle to Kant were preoccupied in one form or another with the issue of the foundations of man's knowledge of natural laws. In another context (that of mathematics) already Plato had grappled (in the *Meno*) with the question of how man can know truths that are necessary and universal, seeing that experience inevitably deals with the actual and particular. Our problem is just exactly this, transported from a mathematical to a physical setting.

13. For a cogent attack on the view that laws can be established by induction see K. R. Popper, *The Logic of Scientific Discovery* (London, 1959), chap. III and New Appendix 10.

The basic fact of the matter—and it is a fact whose importance cannot be overemphasized—is that the elements of nomic necessity and hypothetical force are not to be extracted from the evidence. They are not *discovered* on some basis of observed fact at all; they are *supplied*. The realm of hypothetical counterfact is inaccessible to observational or experimental explanation.[14] Lawfulness is not found in or extracted from the evidence, it is superadded to it. *Lawfulness is a matter of imputation.* When an empirical generalization is designated as a law, this epistemological status is *imputed* to it. Lawfulness is something which a generalization could not *in principle* earn entirely on the basis of warrant by the empirical facts. Men impute lawfulness to certain generalizations by according to them a particular role in the epistemological scheme of things, being prepared to use them in special ways in inferential contexts (particularly hypothetical contexts), and the like.

When one looks at the explicit formulation of the overt *content* of a law all one finds is a certain generalization. Its lawfulness is not a part of what the law asserts at all; it is nowhere to be seen in its overtly expressed content as a generalization. Lawfulness is not a matter of what the generalizations *says*, but a matter of *how it is to be used*. By being prepared to put it to certain kinds of uses in modal and hypothetical contexts, it is *we*, the users, who accord to a generalization its lawful status thus endowing it with nomological necessity and hypothetical force. Lawfulness is thus not a matter of the assertive content of a generalization, but of its epistemic status, as determined by the ways in which it is deployed in its applications.

This approach to lawfulness as imputed rests on a concept of the nature of scientific laws to which more explicit articulation

---

14. It is obviously naïve to think that one can settle the question of the *counterfactual* application "If Caesar's chariot had been a satellite in orbit about the earth it would have moved according to Kepler's laws" by increasing the domain of *actual* applications of Kepler's laws by putting more spacecraft into orbit.

must be given. Present-day philosophers of science have concentrated their attention primarily upon two aspects of "laws": (1) their *assertive characteristics,* having to do with the machinery deployed in their formulation (they must be universal generalizations, must make no explicit reference to time, must contain no overt spatial delimitation, and should be "simpler" than equally eligible alternatives), and (2) their *evidential status*, having to do with their supporting data (they must have no known counter-instances, should be supported by an ample body of confirming evidence).[15] To considerations of this sort one may add yet a third factor, which could be put under the heading of an appropriate *epistemic commitment* having to do with the extent to which we are committed to retention of the law in the fact of putative discordant considerations of a strictly hypothetical character, and thus not of an evidential sort—for this would lead back to item (2)—arising in a choice between it and other items of "our knowledge." The appropriateness of such epistemic commitment revolves about questions of the type: "To what extent is the 'law' at issue justifiably regarded as immune to rejection in the face of hypothetical considerations?" "How should this generalization fare if (*per improbabile*) a choice were forced upon us between it and other laws we also accept?" "How critical is it that the law be true; how serious a matter would it be were the law to prove false?"

This third factor represents an aspect of laws *crucially important to their status as laws.* For no matter what the structure of a generalization might be, or how well established it is by the known data, its acceptance as a law demands some accommodation of it within the "system" of knowledge. Any "law" occupies a place that is more or less fundamental within the general architectonic of our knowledge about the world; its epistemic status is a matter not only of *its own* form and *its own* evidential

---

15. See, for example, the excellent discussion in chap. 4 of Ernest Nagel's book on *The Structure of Science* (New York, 1961).

support, but of *its placement within the woof and warp of the fabric comprising it together with other cognate laws of nature*. The standing accorded to it within the overall framework of our knowledge reflects our "epistemic commitment" to the law, which is thus a matter not of the individual characteristics of the "law" viewed (insofar as possible) in isolation, but of its interconnections with and its relative epistemic embedding among other laws to which we are also committed. We must *decide* upon the epistemic status or ranking of the law with respect to others, and this decision, although in part *guided* by evidential considerations, is by no means *determined* by them alone, but is a matter of the entire range of systematic and epistemic considerations, among which evidential considerations are only one (though to be sure a prominent) factor.

Whereas it is, of course, "we" who "decide" upon the placement of a law in the epistemological scheme of things, and "we" who "make an epistemic commitment" to the law, the crucial point is that this be done on the basis of rational grounds (of complex and varied character) and not on the basis of a merely random choice or personal predilection. The appropriateness of epistemic commitment to laws is therefore not a matter of psychology or of the sociology of scientific knowledge. It is necessary and proper to distinguish between being *in fact* committed to accepting a generalization as a "law" of more or less fundamental status upon the one hand, and being *properly* or *warrantedly* committed to it, upon the other.

In saying that the necessity and hypotheticality of lawfulness are matters of imputation, there is no intention whatsoever of suggesting that the issue is one of indifferent conventions or arbitrary decisions. The imputation is, to be sure, an overt rational step for which a decision is required. But to be justified this decision must be based upon a rational warrant. It must have a grounding in (1) the *empirical evidence* for the generalization at issue in the law and (2) the *theoretical context* of the generalization. Such grounding is required to provide the necessary *warrant* to justify an imputation of lawfulness. Since an element of imputation is involved, laws are not just discovered; they are, strictly speaking, made. This is not, of course, to say they are made arbitrarily. Although they cannot be extracted from the empirical evidence, they must never violate it. Such conformity with "the observed facts" is a key factor of that complex that bears the rubric of *well-foundedness*. Our conception of the origin of the key requisites for a law (nomic necessity and fact-transcending hypotheticality) can thus be summarized in the slogan, *Lawfulness is the product of the well-founded imputation to empirical generalizations of nomic necessity and hypothetical force*.

We must pursue somewhat further this key theme of the warrant for imputations of lawfulness, which we have held to be a question of evidence and of theoretical context. The matter of evidence for scientific generalizations may at this time of day be supposed relatively familiar to the reader. Even the most elementary discussions of scientific method devote considerable attention to the issue of the evidence needed for scientific laws. On the other hand, the bearing of the theoretical context of an empirical generalization in establishing its claims to lawfulness is a perhaps much less familiar factor.

It is not for nothing that branches of science are called *bodies* of knowledge. Scientific knowledge has a complex and highly articulated structure. The laws comprising this structure rarely if ever stand isolated and alone: they are part of a fabric whose threads run off to intertwine with other laws. Scientific laws do not stand in splendid isolation; they interlink with others in the complex logical networks commonly called *theories*. To say this is not to deny that there can be such things as "merely empirical generalizations"—universal propositions which, though well confirmed by the empirical evidence, wholly lack a footing within some ramified theoretical framework. Kepler's laws of planetary motion, Galileo's law of falling bodies, and Boyle's gas law, for example, were all in their day well established and generally accepted prior to securing the grounding provided by a foundation upon some adequate theory. But an aggregation of well-confirmed empirical generalizations does not constitute a science. A science is not a catalogue of observed regularities. It requires a certain *rational architectonic*, relating a variety of empirical generalizations in a rational structure that exhibits their conceptual relevance and their explanatory interconnections. A well-established generalization qualifies as a *scientific law* (in the proper sense of the term) only when it finds its theoretical home within some scientific discipline or branch of science.

Yet, as we have seen, the conditions that establish a generalization as *lawlike* — that is as rationally *qualified* for an imputation of lawfulness on the basis of the usual methodological considerations — do not suffice to *establish* it as a law, because its acceptance as a law involves the claim of lawfulness, and the content of this claim extends well beyond the basis upon which it is justified. To class a generalization as *lawlike* is to say it is a candidate-law on the basis of factual considerations, but to class it as *lawful* is to step beyond this claim into the realm of nomic necessity and hypothetical force.

Various writers have long argued that the very idea of lawfulness is at bottom anthropomorphic.[16] The basic idea is that lawful phenomena are *rule-governed*; the conception of operative rules is the foundation of lawfulness. But this idea ultimately originates in man's first-hand experience of the rules of his social group: rules of behavior, of speech, of dress, and the like. The "pressure" of social rules and the associated sanctions is something of which each man is conscious in his own mind. The rule-conformity of this social context is projected into inorganic nature to provide the concept of lawfulness, analogizing the regularity of social phenomena to those of inert nature, and correspondingly analogiz-

16. For a discussion of historical issues see Edgar Zilsel, "The Genesis of the Concept of Physical Law," *The Philosophical Review*, vol. 51 (1942), pp. 3–14.

ing the alternative range of the socially permissible to that of the naturally possible, and that of the socially obligatory to that of the naturally necessary. In thus viewing man's first-hand experience of social rules as the foundation for his projection of nomic force into the laws of nature one sees the subjective tendency of mind as the model of lawfulness, the ultimate source for the imputation of nomic necessity that is the touchstone of laws.

Thus in saying that laws are man-made — that they result from a human act of according a certain status to specific generalizations — we do not intend to turn our back upon the findings of methodologists of science and theorists of inductive logic. Insofar as their findings conform to the actualities of scientific practice we accept them in full. The doctrine of lawfulness as imputation comes not to negate the standard theory of scientific method, but to fulfil it. We are not attempting a quixotic substitution of "free decision" for scientific method. But we regard the principles of the theory of scientific method from our own perspective — not as procedures for the *establishment* of generalizations as lawful, but as procedures for providing endeavors a rational warrant for *imputations* of lawfulness.

Returning to the idea of lawfulness as a *well-founded imputation*, we remark that both of these two factors, the factual element of well-foundedness and the decisional element of imputation, are necessary to laws. Well-founding is essential because the very spirit of the scientific enterprise demands reliance only upon tested generalizations that have a solid observational or experimental basis. But the element of imputation is also essential. We can only observe what *is*, i.e., what forms part of the realm of the actual, not what corresponds to the modally necessary or the hypothetically possible. The nomic necessity and hypothetical force characteristic of lawfulness thus represent factors that a generalization can never earn for itself on the basis of observational or experimental evidence alone. It has to be *endowed* with these factors.

3. Lawfulness as Mind-Dependent

Our conception of the nature of lawfulness carries Kant's Copernican revolution one step further. Hume maintained that faithfulness to the realities of human experience requires us to admit that we cannot find nomic necessity in nature. Kant replied that necessity does indeed not reside in observed nature but in the mind of man, which projects lawfulness into nature in consequence of features indigenous to the workings of the human intellect.[17] Our view of the matter agrees with Hume's that lawfulness is not an observable characteristic of nature, and it agrees with Kant that it is a matter of man's projection. But we do not regard this projection as the result of the (in suitable circumstances) inevitable working of the epistemological faculty-structure of the human mind. Rather, we regard it as a matter of *warranted decision*, a deliberate man-made imputation effected in the setting of a particular conceptual scheme regarding the nature of explanatory understanding. We thus arrive at a position that is Kantian with a difference. Kant finds the source of lawfulness in the way in which the mind inherently works. We find its source in the conceptual apparatus that we in fact deploy for explanatory purposes: As we see it, lawfulness demands an imputational step

17. A thread running constant throughout the history of philosophy is the thesis that there would be no laws if there were no lawgiver; that the universe would not be intelligible by man if it were not the product of a creative intelligence. We find this theme in Plato's *Timaeus*, in the cosmological argument of St. Thomas Aquinas and the schoolmen, in Descartes and Leibniz, in Butler's *Analogy* and the tradition of natural theology in England. Leibniz puts the matter cogently and succinctly:

> . . . the final analysis of the laws of nature leads us to the most sublime principles of order and perfection, which indicate that the universe is the effect of a universal intelligent power. (G. W. Leibniz, *Philosophical Papers and Letters*, ed. by L. E. Loemker, vol. II (Chicago, 1956), pp. 777–778.)

Kant in effect agrees with the underlying thesis that the intelligibility and rationality of the universe must be the work of an intelligent and rational mind, but shifts the application of the principle from the *creator* of the natural universe to the *observer* of it.

made in the context of a certain concept of explanation. Both of these divergent views agree, however, in making lawfulness fundamentally mind-dependent.

On such a view, laws—even natural laws—are man-made. Does it follow from this position that if there were no men, or rather no rational minds, that there would be no natural laws? Are we driven to a law-idealism as the logical terminus of the line of thought we have been tracing out? The answer to these questions, I believe, must be *Yes*. Given the concept of a law that we actually deploy in these discussions, hypothetical force (and so nomic necessity) is an essential feature of a "law." The mode of lawfulness built into the very concept of a natural law involves an essential reference to the domain of supposition and counterfact, to the hypothetical realm of "what would happen if." And if rational minds were abolished, the realm of supposition and counterfact would be abolished too, and with it lawfulness as we conceive of it—which involves an essential reference to counterfact—would also have to vanish.

The issue of mind-dependency can be clarified by an analogy. One can characterize an object's surface as round or as round-seeming, the foliage of a plant as dense or as dense-appearing, a creature as bird or bird-like, a fabric as silken or as silky (i.e., silk-appearing). In each of these pairs, the former member is not mind-dependent (at any rate not in any overt way), whereas the second member is on the very surface of it mind-dependent, since it deals quite blatantly with how the object at issue strikes the percipient, and thus introduces a reactive mind into the picture. Of course, the mind-dependency at issue in these particular examples is perceptual, but this feature is by no means necessary in general. Thus in characterizing the configuration of rocks at Stonehenge as complex, I endow this arrangement with no perceptual characteristic whatever, but with the mind-dependent *intellectual* (rather than perceptual) feature of being difficult to understand and explain, etc. It is, of course, this intellectual mode of mind-dependency that is at issue in regard to lawfulness. To characterize a generalization as lawful—like characterizing it as fundamental or important or interesting—is to give it a position within an epistemological framework in a way that reflects the stance taken toward this generalization by a comprehending intelligence.

I am not saying simply that laws, being formulated as propositions, presuppose language, and therefore presuppose minds. Nor do I want to go off into Bishop Berkeley's forest. For we are not at the moment concerned with the general idealist position that properties in general require minds. We recognize and admit —indeed regard as crucial—the distinction between the *attribution* of a property to an object by someone (which evidently requires a mind), and the *possession* of the property by the object (which is, or may be supposed to be, an "objective," mind-independent fact). But hypothetically counterfactual propositions are unavoidably mind-related: The hypothetical cannot just "objectively be" the case, but must be hypothesized, or imagined, or assumed. Unlike real facts, hypothetical ones lack, *ex hypothesi*, that objective foundation in the existential order which alone could render them independent of minds.

Of course, in a trivial sense everything that is in fact actually discussed by someone bears *some* relationship to a mind. Unquestionably, no matter what truth *we* may think of, *somebody* thinks of; but what people think is not the crux. Being thought of is not essential to the truthfulness of a truth. And this way of approaching the matter—with reference to what "is thought" to be the case—loses sight of the key issue of *laws*. This would trivialize the issue: It fails to differentiate laws from generalizations that are not lawful, and so the point loses any specific relevancy to lawfulness as such. But just this reference to lawfulness is the essential thing. I have no desire to question the distinction between a fact, say that the cat is on the mat which could continue unchanged in a world devoid of intelligence), and the thought or statement of a fact (which could not). My point is that the *claim* of lawfulness, unlike the *claim* of factuality, involves something (viz., a reference to the hypothetical) that would be infeasible in the face of a postulated absence of minds. A generalization like "All cats are vertebrates (i.e., have backbones)" *if not taken to formulate a law* makes a claim whose correctness is doubtless unaffected if we postulate a mindless universe. But if the generalization is construed in a lawful sense, as asserting that cats *have to* have backbones, with some sort of nomic necessity the story is quite different. For lawfulness "lies in the eyes of the beholder," since the lawfulness of a generalization consists in its being regarded and treated and classified and used in a certain way. All this is impossible in a mindless world. Kant was quite right. Lawfulness is not something that one can meaningfully postulate objectively of a mindless world; it is a mode of "appearing to a mind." For if the hypothetical element (which is clearly accessible only in a world endowed with minds) were *aufgehoben* (annihilated), lawfulness would be *aufgehoben* too. Of course *we* can think of an "alternative possible world" that is unpopulated, and so mindless, but yet lawful—so long as we do not imagine *ourselves* too wholly out of the picture so as to create a genuinely mindless universe. But if we rigorously put aside all reference to the mental—even tacit reference— then the hypothetically possible is lost, and lawfulness is lost with it.

It should be stressed that the hypothetical element at issue here extends well beyond the sphere of overt laws, to encompass all dispositional predicates such as "soluble," "malleable," "fragile," as well as cognate dispositionally classificatory nouns such as "conductor" (of electricity), "nonconductor," and the like. For these all have an intrinsic hypothetical element. The cube of sugar is *soluble* because "*if* it is immersed in water for a sufficient period of time, *then* . . ."; the copper wire is a conductor because "*if* an electric charge is placed at one extreme, *then*. . . ." Applications of all such dispositional nouns and predicates are implicitly lawful. Thus our analysis of the implications of lawfulness are operative here also. Thus insofar as lawfulness is mind-dependent, so is the applicability of such intrinsically hypothetical dispositional qualifiers.[18]

At this point, however, the distinction between laws and regularities becomes important. No doubt nature is in various respects regular; it would take a bold act of rashness to deny that! And this regularity of nature in various respects is no doubt an ontological fact that would remain unaltered in the face of any hypothetical removal of rational minds from within its purview. But the idea of a law involves, as we saw, more than just factual regularity as such, since lawfulness is bound up with nomic necessity and hypothetical force. To say that these factors do not represent objective facts but result from man-made imputations is not to gainsay the objective reality of regularities in nature. Rather, it is to recognize that laws play a role in our conceptual schema that imposes requirements going beyond mere regularity. It is not the regularity claimed by a law but the lawfulness it builds into this claim that is mind-dependent. (The "idealistic" aspect of our law-idealism is thus a qualified one.)

The point can be brought home by means of considerations already alluded to. The thesis that "Oak trees are deciduous" may well represent a regularity that continues unchanged in a mindless world. But the thesis that "Oak trees *have to* be deciduous," in a sense that warrants "If that pine were an oak then it would be deciduous," would not be unchanged. It does not deal with the "objective fact" of regularity alone but brings in a realm of the hypothetically possible—and this, by the very nature of the hypothetical, is vulnerable to a supposition of mind-

18. Someone might content that ordinary, overtly nondispositional predicates like "has a length of one meter" are all covertly dispositional in the manner of "If one were to take a meter rod and lay it alongside, then . . ." This assimilation of all physical predicates to the dispositional ones, though superficially plausible, is at root indefensible. To be sure, all such descriptive predicates have a regularity component and a lawfulness component; but with ordinary "descriptive" predicates the regularity component predominates, while with dispositional predicates the lawfulness component is predominant. The straightforward assimilation of the two cases is thus an unjustified oversimplification.

lessness. Regularity calls for no more than a universal generalization of the type

(1) All X is Y.

But lawfulness, as we have said, goes beyond this to stipulations
of the form

(2) All X is Y, and if z (which isn't an X) were an X then z
would be a Y.

Now (1) is simply an issue of ontological fact: If it is in fact the
case that *All (pure) mercury solidifies at $-38.87°C$*, then this
circumstance could continue operative even if there were no
minds around to think about such matters. But a type (2) generalization—a lawful generalization—would fare differently. Its
reference to z's that aren't X's having to have certain characteristics if they were X's involves claims outside the domain of
ontological fact, claims that make sense only under the (implicit)
assumption that there are minds somewhere upon the scene,
capable of entertaining hypotheses regarding what-would-be-if.
In a mindless universe, the whole domain of the hypothetical
can find no foothold. This, then, is the foundation for our thesis
that laws, involving as they do essentially hypothetical claims,
are mind-dependent in a way that endows them with an inevitable
man-made component.

The critical point that the realm of the hypothetical is mind-
dependent should be argued explicitly. Somewhat reluctantly we
must thus enter at least briefly upon a metaphysical digression
regarding the ontology of the possible. The argument for the
mind-dependency of hypothetical possibilities proceeds as
follows:

1. The natural world comprises only the actual. This world
   does not contain a region where nonexistent or unactualized possibilities somehow "exist." Unactualized hypothetical possibilities do not exist in the world of objective
   reality at all.

2. Nor do unactualized possibilities somehow exist in some Platonic realm of world-independent reality.

3. The very foundation for the distinction between something actual and something merely hypothetically possible is lacking in a "mindless" world. Unactualized hypothetical possibilities can be said to "exist" only insofar as they are *conceived* or *thought of* or *hypothesized*, and the like. For such a possibility to be (*esse*) is to be conceived (*concipiendi*).[19]

In such a way, then, one can argue a denial that possibilities exist in some self-subsisting realm that is "independent of the mind." Inorganic nature — subrational nature generally — encompasses only the actual: The domain of the possible is the creation of intelligent organisms. A "robust realism of physical objects" is all very well, but it just will not plausibly extend into the area of the hypothetical. It would be foolish (or philosophically perverse) to deny the thesis:

> *This (real) stone I am looking at would exist even if nobody saw it.*

But we cannot reason by analogy to support the thesis:

> *This imaginary stone I am thinking of would exist even if nobody imagined it.*

The objectivity of the real world does not underwrite that of the sphere of hypothetical possibility. This sphere is mind-dependent, and so consequently those intelligent resources which, like laws, are hinged upon it.

Of course, Idealists of the old school (F. H. Bradley, B. Blanshard) would not accept our idea of laws as a *via media* between mere regularities on the one hand and the logically

19. To say this is not to drop the usual distinction between a thought and its object. If I imagine this orange to be an apple, I imagine it *as an apple* and not as an *imaginary* apple. But this does not gainsay the fact that the apple at issue *is* an imaginary apple that "exists only in my imagination."

necessary on the other. Indeed their reasoning seems to proceed by elimination: There is no such intermediary and natural laws are not mere regularities, *ergo* they are truths of logic. Now, of course, one would incline to counter this position with the claim that there is nothing inherently self-contradictory about denying the laws of nature, whereas there is something self-contradictory about denying a logical truth. The old-line idealists will respond that law-denials will turn out to be self-contradictory once we have learned enough about the *system* of which they are a part. (It is certainly possible to hold a belief on grounds one believes to be empirical but which later turn out to be logically necessary.) They hold that, when science is complete, all lawful relationships will be revealed as somehow logically necessary. This (Blanshard's) position is akin to that of Leibniz: All empirical propositions are analytic although they may not appear so to our finite minds. This position is not patently untenable, but it places a burden on one's faith in ultimates that has little to recommend it.

The key points of the argument developed in these three sections can be summarized as follows:

1. The concept of scientific explanation is such as to require *lawfulness* in the generalizations employed.
2. Lawfulness requires the factors of nomic necessity and hypothetical force.
3. Nomic necessity and hypothetical force both in significant measure go beyond the sphere of what can be established by observation and experiment.
4. Lawfulness can thus never be wholly based upon an observational foundation. Rather, it represents an *imputation* that is (or should be) well founded upon evidential grounds. (The key factors in this well-foundedness are the *correspondence-to-fact* aspect of empirical evidence and the *systematic-coherence* of filling the generalization into a fabric of others that in the aggregate constitute a rational structure, an integrated body of knowledge that constitutes a "branch of science.")

5. Laws are therefore in significant respects not discovered but made. A law, unlike a simple assertion of regularity, involves claims (viz., of nomic necessity and hypothetical force) that are mind-dependent and cannot be rested simply upon objective matters of observed fact.

6. Our position thus has the character of a qualified idealism. Lawfulness is not *just* a matter of the observable facts but involves, through reference to the factors of nomic necessity and hypothetical force, an essential element of transfactual imputation, and thus is in a crucial respect mind-dependent.[20]

4. Causal Laws: The Principle of Causality and Its Limitations

Causal explanation is unquestionably the most prominent of all modes of scientific explanation, and any discussion of explanation would remain grossly incomplete if it failed to deal with the idea of causality. The best approach to the concept of a cause is through the correlative concept of a *causal law*. A causal law is one that corresponds to empirical generalization of the type: "Whenever conditions of type X are realized, a phenomenon of type Y will be present." On the traditional conception of the matter, a type X event is *"a cause"* of type Y only if, whenever a type X event is realized,[21] a type Y event invariably succeeds or coexists with it. An example of a causal law of this sort is represented by the law of falling bodies initially enunciated by Galileo:

Whenever an object has been released from altitude for free fall *in vacuo* for a period of t seconds, it will move toward the earth's surface with a velocity of $\frac{1}{2}gt^2$.

20. Some of the ideas dealt with in the section are discussed in a very illuminating way in A. C. Ewing, *Idealism: A Critical Survey* (London, 1934, 3rd ed., 1961), see chap. viii, "Idealistic Metaphysics."

21. Or, more fully, "is realized under appropriate circumstances" — which then need to be spelled out.

A *causal explanation* of an event is, correspondingly, an explanation of its occurrence in terms of its causes, and calls for the specification of the actual conditions and circumstances that underwrite its explanation in the context of specified causal laws. One (rather oversimplified) example would go as follows:

Q: Why is that object now falling with a velocity of $8g$ — i.e., what are the causes of this?

A: It was released *in vacuo* from altitude four seconds ago, has since been moving in free fall, and *Whenever an object has been released....*

At its very simplest, a causal explanation is a direct deductive subsumption argument of this sort, placing given occurrences within the scope of a causal law.

In general, of course, the situation is more complex. The causal background of most occurrences is highly variegated. A vast multiplicity of factors may be operative in producing the event; the bankruptcy of a manufacturing firm may be due in part to a lessened demand for its products, in part to a rise in price of the input materials its products require, in part to the failure of a major creditor to discharge his obligation. A whole host of causal circumstances may "conspire" in producing the caused event, and many causal laws may be jointly operative in its production. We would then speak of "*a* cause" in connection with one of these causal factors; i.e., as one of the elements in the complex causal picture, and characterization as "*the* cause" would only be applicable to the aggregate sum-total of all these factors.

It follows from this recognition of complex, multi-constituent causation that to say of an occurrence that it is causally explicable is not to claim that this event falls squarely under the purview of some one causal law. It would suffice that the event can be accounted for within the framework of causal laws in general, though possibly requiring recourse to a plurality of laws for its explanation. To be causally explicable, an event need not fall within the scope of one single causal law but need only be covered by the entire fabric of the totality of causal laws.

The idea of a *cause* is correlative with that of a causal law. One can speak of "causes" only where cause-effect relationships are held to be operative, and to hold this is to invoke the workings of causal laws. The step from the operative causes to the caused results cannot be made without the mediation of causal laws. Causal laws thus provide the indispensable setting within which the notion of cause-effect can be deployed and without which the idea of causality would remain inapplicable. The ideas of causality, of causal laws, and of causal explanations are inseparably intertwined correlatives: Each can be brought to bear only where the others are applicable.

There is no need to dwell at length on the important place that the concept of causation has both in the commonplace affairs of everyday life and in the more rigorous context of various branches of science, which, like history, astronomy, evolutionary biology, or forensic medicine, are extensively concerned with the scientific explanation of particular, concrete events. The idea of causality — and the mechanism of causal explanation that goes with it — is certainly among the most prominent and pervasive concepts with which we operate throughout all attempts to understand the world we live in.

Due recognition of this supremely important place of the conception of causality throughout man's attempts to understand his environment has led philosophers to espouse the "Principle of Causality" — the thesis that *Every event has a cause*. In its full articulation this thesis embodies the claim that any and every event E can be subsumed under a complex of causal laws of the type: Whenever C-type circumstances are realized, an E-type result will ensue.

This thesis that all events can be accounted for within the framework of causal law has far-reaching implications for the topic of our present discussion. It would lead to the immediate consequence that any and every event can be explained causally, that there are no limits to the range of applicability of causal explanation within the sphere of natural events. The question of the correctness of the Principle of Causality is thus of central importance for the theory of explanation.

It is clear upon even cursory reflection that the Principle of Causality cannot be an *a priori* truth: It makes an empirically refutable claim about the contingent structure of the course of nature. It is thus necessary to regard it as an (extraordinarily broad) empirical generalization that may be more or less well supported by the evidence thus far in hand. Such, clearly, must be the status of a doctrine that asserts:

> Every event that occurs belongs to a class E all members of which are invariably preceded (or accompanied) by events of an associated class C.[22]

That this thesis is thus empirically vulnerable is patent on the very surface; it makes the factual claim that *every event invariably has a certain feature* of association with other events. The claim at issue is that the world is of a certain by no means inevitable type—that occurrences in it have a particular pattern of orderliness. This is certainly no logically necessary feature of any and every possible world. Even if it should turn out to be an actual feature of this world (which is something we shall call into question), it would have to be a contingent feature that cannot be claimed to obtain on the basis of *a priori* considerations but only on the basis of the factual evidence one can amass. It is thus clear that such a principle must be viewed in the light of a descriptive, empirical thesis, albeit one that is very fundamental in its nature and very comprehensive in its scope. As elsewhere, then, we

---

22. The machinery for specifying such classes E and C must be restricted in appropriate ways if this thesis is not to be trivialized. Such complexities need not concern us at this present point. Moreover the link between E and C must not be a *logical* one; they must be logically independent. Thus an aircraft's "breaking the sound barrier" can be advanced as the cause of a "thundering noise" but cannot be regarded as the cause of a "sonic boom," because "sonic boom" is *defined* as the noise made by an aircraft breaking the sound barrier, and so the requirement of logical independence would be violated.

must look to the extent to which the generalization is supported by the evidence in hand. Viewed in this perspective, it soon becomes clear that the Principle of Causality is in fact false, or at any rate very likely false in the face of present-day science.

5.   The Stochastic Revolution and the Downfall of Causality

Nowadays it is a commonplace observation that the entry of irreducibly stochastic laws into natural science via physics—that paradigm of scientific respectability—has made a revolutionary impact upon our *concept of nature*. Even schoolboys have heard something of the quantum theory and its dethronement of the deterministic conceptions of classical physics. It has also been argued, in discussions that have perhaps generated more heat than light, that the statisticization of microphysics holds vast implications for our *concept of man*, because of purported bearings upon the age-old problem of free will *versus* determinism in human affairs. In the hue and cry over these two highly exciting phases of the stochastic revolution, a third, less striking but perhaps no less significant effect has been relatively under-emphasized—the necessity for basic revisions in the concept of scientific explanation itself. A close look at this development is essential to our scrutiny of the nature of explanation.

One of the central themes of modern physics relates to the existence of *random* events represented by such irreducibly stochastic processes as radioactive decay.[23] In these contexts there are specific events—such as the decay of an unstable atom

23. I speak of "*irreducibly* stochastic processes" here because, of course, classical deterministic physics too allows probabilistic considerations to enter in. Thus the range of probabilistic theories in general includes also the old classical statistical theories (such as, for example, the kinetic theory of gasses) which are in principle deterministic, that is—where the "indeterminacy" is only a result of an incomplete description of the initial state of the system. Brownian motion, for example, is considered to be part of classical deterministic physics. Although the usual treatment of Brownian motion is statistical, a strictly deterministic treatment is in principle possible.

of some very heavy element — which, according to the standard view of currently accepted scientific theories, take place without the operation of any "causes," and with respect to which causal explanations cannot be given. Indeed, as we saw in the discussion of probabilistic discrete state systems, there can be natural systems whose history includes *not a single caused occurrence*; that is, none of whose states is "caused" by its preceding states in the sense of causality specified at the outset. Events of this type make manifest the falsity of the Principle of Causality, or at any rate its probable falsity, allowing for potential imperfections in the present state of scientific knowledge of the matter.

The downfall of causality has profound implications for our understanding of the nature of explanation. Only gradually have philosophers of science, students of scientific method, and scientists concerned about conceptual fundamentals come to occupy themselves with something whose necessity they have granted but grudgingly — a fundamental re-examination of the very meaning of *explanation*. Heretofore causal explanation has been the standard mode of explanation, and thus the model has been a mode of explanation that consists in the *deductive* derivation of an explanatory conclusion from *universal*-law premisses. But in the wake of the dethronement of causality in quantum physics the deductive conception of explanation suffered a hard blow — indeed a mortal one so far as its claims to exclusiveness are concerned.

According so naturally as it does with the mechanistic approach of classical Newtonian physics, the deductive conception of explanation rapidly established itself after the beginnings of methodological consciousness in the second quarter of the nineteenth century. Throughout the century, it was adopted as the correct theory of scientific explanation. Such illustrious thinkers as Comte, Whewell, Mill, Jevons — in short, most major nineteenth-century theorists of scientific method — all accepted this conception of the nature of the explanation.

So deep-rooted did this way of looking at explanation come to be, that the stochastic revolution of turn-of-the-century physics seemed for a long time to leave it to all appearances unaffected. Various important twentieth-century students of science, including Rudolf Carnap, Norman Campbell, Karl R. Popper, and Herbert Feigl gave continued and powerful support to the deductive view of explanation. Perhaps its most elaborate and painstaking formulation was given as late as 1948, in an important article by C. G. Hempel and P. Oppenheim, where however there is an explicit, although unemphatic, recognition of its limitations. Adherence to the deductive conception of explanation continued well into the present century.[24]

Only since the 1940's, with the fading influence of logical positivism, so heavily imbued with nineteenth-century conceptions, have statistical explanations come to be recognized as deserving not only a measure of acceptance but almost a place of prominence. At last a state of theoretical affairs was realized which developments in physics since the turn of the century rendered appropriate and indeed inevitable. (Not that die-hard support of the exclusivity of the deductive concept of explanation cannot be found at the present writing!) However, a dogmatic insistence that only deductive accounts are to qualify as explanations flies in the face of historical experience. At various junctures, scientists have been inclined to insist that explanations cannot transgress certain limits. The sphere of proscribed explanatory instrumentalities has included such items as action at a distance, creation *ex nihilo*, and noncausal processes. Because

24. See for example, the article by May Brodbeck "Explanation, Prediction and 'Imperfect' Knowledge," *Minnesota Studies in the Philosophy of Science* ed. by H. Feigl and G. Maxwell, vol. 3 (Minneapolis, 1962), pp. 231–272. Here the problem of explanations on the basis of statistical laws is resolved by the (undefended) declaration that "from statistical generalizations, we do not [Is "cannot" intended?—N. R.] deduce 'with probability' that a certain event will occur, rather we deduce exactly the relative frequency or 'probability' with which an event will occur in a certain group." This, of course, puts stochastic laws *hors de combat* in any discussion of particular occurrences.

of the repeated failure of such limiting restrictions, there has been a marked decline of *a priorism* in regard to scientific explanation: in the face of modern developments in science, people are increasingly disinclined to say that an explanation cannot proceed this or that way. (Even in psychology there is now a notably lessened tendency to dismiss parapsychological findings out of hand as a matter of "general principles.")

Various authors still preach on the Kantian text that the Principle of Causality is *regulative* for scientific inquiry in the sense that a failure to accept the principle would constitute a somehow defeatist attitude toward the possibilities of arriving at scientific understanding of the world about us, an attitude that might hamper or even block success in scientific work.[25] Construed in this way, the principle *is not a thesis but a rule of procedure*, an operational maxim enjoining the search for causes. Now an operation maxim may well be highly serviceable even if it is a "useful fiction" based on a false premiss: "Always carry forward your research as though the events at issue were caused" — "Always drive as though all the other motorists were incompetent." A useful principle of operation need not correspond to a true doctrinal precept, there being a vast difference between a procedural *rule* (Treat all men as your friends) and the corresponding factual *thesis* (All men are your friends). The former may well be maintained in the face of counterinstances to the latter.

Thus the Principle of Causality could well prove to be useful as a regulative principle even though false as a thesis. But even this does not seem to be the case, and we can spare ourselves the intellectual inconvenience of a "useful fiction." For the fact is that it is by no means unavoidably detrimental to the aims of science to demote causes from their traditional pedestal of the

25. A forceful presentation of this view — given, however, without endorsement — is presented in John Hospers, *An Introduction to Philosophical Analysis* (2nd ed., Englewood Cliffs, N.J., 1967), pp. 317–318. This book contains an elementary but very cogent and lucid discussion of the principle of causality.

primary target of inquiry. As we observed at the outset, causes are correlative with causal laws, and what is crucial here is not the *causality* but the *legality*. It is, after all, the discovery of the laws of things, regardless of whether they turn out to be universal or probabilistic, that provides a criterion of success in scientific inquiry. Even those natural systems, in the examples from modern, indeterministic physics, in which no caused occurrences take place at all are rigidly governed by specifiable laws, laws to be sure of the probabilistic rather than universalistic type. This yields the important lesson that the fundamental factor in scientific understanding will be a grasp not necessarily of the *causes* but of the *laws* of things. Thus the appropriate "regulative principle," it would seem, is not so much the Principle of Causality ("All events have causes") but a *Principle of Legality* ("All events are law-governed"). We would do well to replace the classic maxim *rerum cognoscere causas* (to know the causes of things) with the version *rerum cognoscere leges* (to know the laws of things). And from this perspective, the de-emphasis of causes as governing objectives of scientific research need not be viewed as a dire impediment in the path of inquiry.

The recognition of stochastic events and its consequent acceptance of probabilistic laws — laws which, not being universal, will not be causal laws — has a decisive role in restricting the realm of causal explanation and thus in forcing a reappraisal of the very concept of causality itself. It forces upon us the recognition that "causally inexplicable" need not mean "mysterious" or "preternatural." There is nothing inherently "strange" or "mystical" in the stochastic processes with which modern science confronts us. Quite the reverse; we are taught the lesson that the operation of "chance" processes in nature can be a matter of prosaic commonplace. All the same, the recognition of the place of stochastic events in the natural scheme of things does effectively invalidate the Principle of Causality as a metaphysical thesis, and in consequence places definite limits about the range of application of the procedure of causal explanation. (And it is only fair to the historically influential concept of causality to stress that the exceptions to the classical Principle of Causality with which modern physics confronts us do belong to certain rather specialized and restricted categories of events.)

All the same, the recognition of this stochastic sector (however small) has important implications of principle, as we have seen. It shows that we must not identify the realm of science with the realm of causality. Science is something too fundamental and deep-rooted not to survive a breakdown of universal causation. Phenomena may fall outside the sphere of causality without thereby moving beyond the horizons of science. Those essentially stochastic processes that lie outside the causal realm, moreover, are not mysterious and irrational. These "chance" phenomena are perfectly "natural" and indeed law-governed. Although such occurrences will violate the classic Principle of Causality they need not be taken to violate the classic Principle of Sufficient Reason; they can be rationalized by being placed in the context of laws in such a way that their occurrence can be "accounted for." An event of this sort may not be "causally explicable" but it nevertheless can be rendered thoroughly *understandable* within the framework of science.

6. Explanation, Scientific Understanding, and the Aims of Science

It has been said that the central, definitive task of science is the specific prediction of the future states of natural systems. On this view, *prediction* represents the ultimate aim of science. But, as we have seen in connection with a variety of examples, there can be perfectly straightforward natural systems of the stochastic type the prediction of whose specific future states is in principle impossible. Consideration of this fact, combined with a realization that the study of such systems is certainly an integral part of the necessary work of science, strongly suggests that it would be mistaken to stipulate that predictability is essential to science.

It might thus seem plausible to take the view, which many writers have indeed taken, that the central, definitive task of science is better construed in terms of explanation rather than prediction, since this latter may be wholly impossible (and impossible not just *de facto* but even in principle) in cases when the former resource lies within reach. But even this position is rendered untenable by the recognition of cases (and we have seen many examples of them) in which "scientific understanding" of a system is perfectly possible even in the total absence of a capacity to *explain* its states.

The fact is that it is altogether wrongheaded to say, as one often hears it said, that the factor of prediction or of explanation or of "control over nature" represents *the aim* of science. Of course it may be to some extent problematic whether someone who speaks in this way intends a reference to the *sole* aim or the *primary* aim. But actually, the thesis in view is incorrect on *either* construction, even the weaker one. For the fact is that *science has a plurality of coordinated objectives.* Science seeks not only to facilitate retrodictive reconstruction of the past and to explain what has happened and what presently goes on, but to predict what will probably go on, and to afford men the instrumentalities of control — or at least partial control — over it. It thus is, or should be, quite clear that science does not have a single, monolithic aim, but a multiplicity of purposes, some basic and theoretical (explanation, prediction, and retrodiction), and others consequent and practical (control).

These various factors have a differential bearing upon the various branches of science and bear upon them with varying weight. In linguistics, for example, the explanatory element is paramount. In medicine we could not rest content with *ex post facto* explanation, no matter how adequate; the pivotal factor is control (and indeed in purely empirical medicine we could attain a reasonable degree of control in the face of machinery inadequate for all explanatory purposes). In meteorology we might, and in cosmology we almost surely would be willing to settle for the capacity to explain and predict, even if our ability to control the phenomena of the subject were nil. The relative importance of explanation, retrodiction, prediction, and control could thus be expected to vary substantially from one branch of science to another.

Viewed in this perspective, science has a multiplicity of functions that can be grouped into two discriminable categories — distinct, yet not unrelated. There is the side of *theoretical* systematization in the exploitation of laws for explanation, prediction, and retrodiction. This theoretical side of science would be of interest to a purely spectatorial intelligence, a disembodied mind that was in no way an active participant in the affairs of the world (regarding which it had some noninteractional source of information). On the other hand, as an active participant in the great drama of the physical universe, man properly seeks through science to acquire a measure of *control* wider than what his prescientific intelligence would otherwise afford him. Viewed as a whole, science is consequently a multipurpose endeavor, having not one monolithic aim but a varied spectrum of related objectives. To speak of explanation (or prediction, or retrodiction or control) as "the aim" (sole or pre-eminent) of science is to fall into an error analogous to that committed by the monolithic *summum bonum* theories of traditional ethics that locate "the aim" of human life uniquely and exclusively in "happiness" or "knowledge" or "piety" or some other single good. It would seem that both approaches are myopic, concentrating upon one factor to the exclusion of other, equally necessary elements of what is a complex, many-sided structure.

Once we proceed beyond the purely manipulative issue of control to the theoretical aspect of science (explanation, prediction, and retrodiction), it would appear that the heart of the matter does not reside in any one of these resources as predominant over the rest but in something common to them all. Each of them, specifically including explanation, may fail us in certain cases in which "scientific understanding" is yet altogether possible. In the cases of the type that are at issue here, and of which various instances have been given, we may certainly have a great deal of information about the systems involved, and can even, in a sense, possess *all* the information about the functioning of the system that can possibly be known. We may thus, for example, know, *inter alia*, the entire history of the functioning of the system, and all of the laws (of all relevant types) governing its behavior. The thing we may be entirely unable to do is to "predict" — or to "explain" — *certain specific occurrences* in any sense answering to any reasonable usage of the term. If we are not to say that scientific knowledge is in principle impossible in such cases, then we must construe scientific knowledge not in terms of an ability to explain (or predict or retrodict) specific states but rather in terms of a grasp of the generalized knowledge of the laws of operation of systems in accordance with which their specific states come to be realized. Thus as far as the theoretical aspect of science is concerned it is *systematization* — that essential knowledge of laws in terms of which explanation and rational retrodiction and prediction all alike become possible — that comes closest to answering the question of "the aim" of science.

The consideration of stochastic systems forces us to the realization that scientific understanding can be present despite an impotence to explain (predict, etc.) *even in principle* certain particular occurrences. As regards explanation, this leads us to recognize the fundamentality of description, being able to deploy laws so as to describe the modes of functioning of natural systems. As regards prediction, it becomes apparent that it is not requisite for scientific understanding to be able concretely to foretell the future, but only that, by invoking natural laws, we be able to bring to bear a concept of possibility and impossibility in terms of which to canalize our expectations, to be able to say what *can* go on.

The root task of science ought thus to be thought of as a fundamentally *descriptive* one: the search for the laws that delineate the functioning of natural processes. For on the one hand, these laws can, as we have seen, operate even in the absence of all prospects for explanation (and prediction or retrodiction). And on the other hand, reliance upon laws is an essential and necessary instrumentality for the accomplishment of the various principal tasks of science (explanation, prediction, retrodiction), indispensable in all those cases where these tasks can possibly be accomplished. It is thus our grasp of natural laws, be they universal or probabilistic — not our capacity to explain, predict, and so on — that appears to be the basic thing in scientific understanding. For it is undeniable that a knowledge of the pertinent laws goes a long way with endowing us with all that we can possibly ask for in the way of an understanding of "the way in which things work" — even in those cases where the specific desiderata of explanation, prediction, etc., are, in the very nature of things, beyond our reach.

Let us place these considerations into the context of the traditional clash between scientific realism and instrumentalism. Realists hold that scientific theories describe the world as it really is. Instrumentalists maintain that science affords us tools for predicting and controlling observed phenomena, without necessarily furnishing any materials for depicting the real world. (Of course, a mode of "realism" is now in view quite different from that at issue in the realism/idealism controversy.) Realists, holding that science can describe physical reality, tend to stress the theoretical aims of science: the description and explanation of what goes on in nature. Instrumentalists, as agnostics regarding the description of physical reality, tend to stress the practical goals of prediction and control. The tendency of the considerations we have just been developing — insofar as they support the fundamentality of the theoretical aims of science vis-à-vis the practical — inclines toward the realistic side of this dispute. But this result of realism *once science is given* will have to be moderated in the light of the instrumentalistic justification of science itself which we shall develop below.

By way of conclusion, o͏
serves brief summary. It i
standing to afford us a previ
to give us the means for e͏
pened. It suffices that we be
realm of possibility" broad
standpoint of the naive, or ͏
"conceivable." The key t͏
capacity to exploit a *know*
standing of the past and to

136

sorts of *problems* t͏
for example, the͏
artifact like a ͏
workings of͏
classify ͏
issue.͏
met

7. Explanatory Frameworks and the Limits of Scientific Explanation

"What sorts of things are candidates for scientific explanation; what is the potential range of the explanatory problems of scientific inquiry?" The answer to this fundamental question would seem to be: "Any and all facts whatsoever." Nothing is in principle placed outside the purview of science. The conceivable subjects of scientific explanation therefore exhibit an enormous, indeed an endless variety. All the properties and states of things, any and all occurrences and events, the behavior and doings of people, in short, every facet of "what goes on in the world," can be regarded as appropriate objects of scientific explanation.

But although science does not exclude anything from its purview, there is perhaps some range of fact that is located outside the effective range of scientific explanation. Certain facts within the field of view of science may yet lie outside the reach of its grasp. Perhaps there are *a priori* theoretical grounds for having to exclude certain facts from the effective scope of science as lying beyond its operative limits. In coping with this issue of the explanatory completeness of science, we must first recognize that there are *alternatives* to modern science as an instrument of explanation.

Explanations can, of course, be classified according to the

which they address themselves; whether,
relate to human actions, the activities of an
atch, or a sector (e.g., the meteorological) of the
inanimate nature. Proceeding in this way one would
xplanations according to the type of explanandum at
But one can also classify explanations according to the
ods that are employed, and more specifically yet, according
o the nature of the sorts of *concepts* that are brought to bear for
explanatory purposes. For example, physical explanations,
astrological explanations, animistic explanations, or numerological explanations can all be put forward to account for one and the
same fact, say, the occurrence of a drought.[26] Here the various
explanations are differentiated according to the different sorts of
concepts brought to bear in the explanans for accomplishing the
explanatory job. In the drought example, these different categories of explanatory concepts would include physico-chemical
interactions, astral influences, occult forces, and others.

It should not be concluded from the character of this specific
example that some one of such alternative methods of explanation
is necessarily and in principle right and the others wrong, or even
that one is necessarily better than the rest. One may think here of
the extensive recent discussions of the explanation of human
actions in terms of *reasons* on the one hand and in terms of *causes*
on the other. It is in principle a perfectly conceivable situation
that such alternative methods of explanation need not be a matter
of better or worse, or greater or lesser adequacy, but may simply
reflect differences of aspect in the facts being dealt with, so that
the issue becomes a *perspectival* one.[27] For an observer to say

26. Note that we are here dealing with dimensionally different explanations
of *the same occurrence*, and not with aspectively different explanations of
various factorial components of a complex phenomenon, such as the political,
military, and economic aspects of a certain foreign policy development. The
crucial difference between factors and dimensions is that dimensions cannot be
combined, whereas factors can combine in complex causation, with each one
contributing its own weight to a composite result.

27. To say this is not, of course, to deny that a certain mode of explanation
can be wholly inappropriate in some cases. The rational-purposive approach

that someone replied to a greeting in the customary way "because of habit" is to give an explanation in different terms from the agent's own statement that he did so "because it was the polite thing to do"; but the correctness of one of these perspectivally different explanations does not exclude that of the other.

The mode of "perspective" now at issue requires a closer examination. Here the idea of an explanatory framework becomes very useful. By an "explanatory framework" we understand a cluster of mutually relevant conceptions in terms of which explanations of a group of related facts regarding natural occurrences could be developed. An explanatory framework is thus a family of basic concepts and principles that furnish the machinery needed for an entire range of applications in the study of some aspect of nature. When a man is taken ill, for example, the onset of his malady might be explained in terms of biomedical causes (e.g., a viral infection) or psychosomatically (e.g., as the result of an auto-suggestion induced by fear of failure) or occultistically (e.g., the "evil eye"). Such explanatory frameworks can, moreover, exist at various levels of generality, as the scientific framework includes the biomedical which in turn includes the immunological. In each case, the explanatory rationale within which the explanation of the occurrence in question proceeds is developed with reference to a certain group of fundamental processes to which explanatory efficacy is imputed (infection, autosuggestion, malign influence, etc.). Within the framework of early 17th-century physics, which proceeded to give all explanations by means of contact-action, the sort of action-at-a-distance at issue in Newton's theory of gravitation was just as heretical as psychokinesis appeared in the explanatory framework of 19th-century science.

One of the most striking prospects opened up by 20th-century physics is that nature is not uniformitarian in regard to size.

---

which is so crucial in the explanation of human actions, for example, may be quite out of place in other contexts, which is why one speaks disparagingly of *anthropomorphism.*

Conceptions as to how nature works at the level of the macro-objects of everyday experience—based on the ideas of space, time, and causality familiar from the observations of ordinary life—may require drastic revision, nay, downright abandonment, at the level of the small-scale microphysical phenomena. The workings of nature may prove to be phenomena of scale: Principles standardly operative at one level may be wholly inapplicable at another. A clear example is afforded by the "slit experiment" of quantum optics, which reveals that the behavior of light is anomalous—light is not a particle and not a wave. Neither of these two models works in all circumstances, and one summary of the situation is that the behavior of light in the small is *conditioned by the environment*, in that only the total system *as a whole* apparently suffices to determine the behavior of light units. At any rate, the prospect thus arises that different explanatory frameworks may be operative in dealing, at different levels of scale, with "the same" sector of nature.

The examples of such occult explanatory frameworks as those of numerology (with its benign ratios), astrology (with its astral influences), and black magic (with its mystic forces) indicate that explanatory frameworks can have diverse degrees of merit. Their critical assessment renders possible a comparative evaluation, or, more radically, even makes it possible to call the legitimacy of an entire explanatory framework into question *en bloc*. For the explanatory efficacy of the whole range of conceptual tools comprising such a framework can certainly be denied in some cases upon highly convincing grounds. We defer until the next section a consideration of the grounds for choice among alternative explanatory frameworks, grounds upon which the superiority of the scientific framework over its alternatives can be established.

Even after one recognizes the pre-eminence of the explanatory framework of science, the question of the *comprehensiveness* of its explanatory range yet remains. Can science actually explain everything? It might seem at first sight that the answer here would have to be a negative one, as indicated by the upshot of an argument as old as Aristotle's *Posterior Analytics*.[28] In brief summary this argument goes as follows: To give an explanation of one fact as explanandum invariably involves the use of others in the explanans. An infinite regress will thus result. This regress can be terminated only if there are certain "ultimate" facts — facts not themselves explained, although they are available for use in the explanation of other facts. Such "ultimate" facts will play the role of basic premises in science much as the axioms are basic in a system of geometry. And these ultimate facts will represent the limits of scientific explanation, for although science uses them in giving explanations, they will themselves lie outside the range of scientific explicability.

This regress argument to an eventual bedrock of "ultimate" facts is seriously defective. This is so because there is no valid reason for postulating that the explanatory regress at issue must be a *terminating* regress that comes to a stop at a particular point of bedrock "ultimate fact." Consider an analogy. Suppose one explains today's occurrences in terms of yesterday's, and yesterday's in terms of those of the day before, and so on. This, quite obviously, is a process one cannot carry through to the finish; one will eventually have to stop someplace. But because one *does* stop at a certain stage it does not follow that one *has to* stop at *this* stage. (Or, at any rate, there is no *reason of principle* why this process could not be interminable, due to time's stretching endlessly into the past *ad infinitum*.) In this setting there will be many (infinitely many) events that are *unexplained*, but there need not be any that are *inexplicable* in the hypothetical sense that we could not explain them even if we would. The process does inevitably stop, but it does not stop inevitably; one is nowhere confronted by an impenetrable barrier. This example helps to show why the (true) premiss that facts must be explained

28. See Bk. I, sect. 3.

from facts does not lead to the conclusion that there must exist a body of ultimate facts.

It is not for any reason of principle that the thesis that "Science can explain everything" must be rejected; it is because modern science itself, in the context of irreducibly stochastic processes, brings to light certain inexplicable facts. The limit of science does not lie at some theoretical boundary of ultimate facts but at the factual boundary of stochastic occurrences. In the light of these, the dictum in view must be revised to "Science can explain everything explicable." We arrive at the (somewhat more modest) thesis that all facts can be explained scientifically that can be explained at all in ways acceptable to a rational mind.

8. The Problem of Explanatory Ultimates

We have refused to recognize a category of "ultimate *facts*." But there is yet another aspect to the question of explanatory ultimates, the issue of ultimate *principles* of explanation. Consider questions of the type traditionally called "metaphysical" which also relate to "ultimates" of a certain sort: *Why does anything exist at all? Why is the nature of existence as it is?* There are quite different explanatory rationales within which an answer to these questions can be attempted: the theological rationale that postulates a creator-god, the mystic rationale that invokes an inscrutable "chance," the naturalistic rationale that looks to some inner exigency of the universe itself.[29] No matter what position one feels inclined to take here, certain basic methodological features of the problem should be noted.

---

29. There is, of course, also possible recourse to the intellectual weariness of a positivism that dismisses these queries as illegitimate questions that "do not arise," and rejects traditional metaphysics as meaningless nonsense. But this intrinsically not unattractive — and certainly convenient — position has always encountered difficulty in showing just why (convenience apart) these issues are to be classed as nonsense.

The questions are, pretty clearly, not ones that can in the final analysis be settled on the basis of evidential considerations of the usual substantive sort. The issues involved are not "internal" but "systematic" to the study of nature, that is, they do not deal with *specific facts* regarding what takes place *within the course of events in nature*, but rather with certain *general principles* operative *with respect to the course of events itself*.

If these questions are not to be dogmatically dismissed as improper, then one must deal with them on the basis of essentially regulative assumptions as to how things work in the world. This line of thought leads back to the concept of an explanatory framework. For such "ultimate questions" can be seen as questions regarding the basic explanatory framework to be deployed in explaining nature's ways. Are we to resolve the issue in essentially causal terms, thus making essential reference to an agency exterior to the universe itself (e.g., God)? Are we to explain in terms of that nonexplanation called "chance"? Is the answer to be sought in some inner exigency of the universe itself? It is only by reference to such framework principles characteristic of the very style of our entire approach to explanation that our "ultimate" questions can be answered, and indeed it is just this that makes them "ultimate."

Explanatory frameworks can, as we have said, be criticized *en bloc*. Such criticism is not, however, possible *a priori*. It is not feasible to establish "on general principles" that numerology, say, or black magic will not work for explanatory purposes. It is perfectly conceivable in the abstract that the ways of the world could be so ordered as to establish the efficacy of these frameworks. The evaluation of explanatory frameworks must proceed *a posteriori*. But what criteria are operative here, by what standards are explanatory frameworks to be evaluated? Since the "choice of a framework" plays such a critical role, we must consider more closely just exactly what is the nature of the *choice* that is at issue. Two considerations are supremely important here. (1) There are genuine alternatives so that there is an authentic choice, but yet (2) this choice is not an arbitrary one to be made with indifference. Both of these points warrant further consideration.

The scientific framework is but one among various alternatives; for example, the occult framework embracing numerology, astrology, black magic, and the like. But wherein does the superiority of the scientific framework reside? Not in any *a priori* factor of *intrinsic* superiority; as already observed, the working of things in this world of ours might well be such as occultism would require. Moreover, this superiority does not in the final analysis obtain on the side of *explanation* at all; a sufficiently sophisticated occultism might well "explain" all the actual occurrences in nature quite as well as science. (Think of the irrefutable efficacy of the concept of "fate" as an instrument for explaining human actions.) The fact is that the superiority of the scientific framework comes about not so much in point of explanation but in point of *prediction* and *control*. It is in *this* regard, not with respect to explanation at all, that the scientific framework is in the final analysis able to make good its claims to predominance. The superiority of the scientific framework of explanation is thus established by essentially practical criteria, and obtains not in principle in a way that is somehow *de jure* and *a priori*, but in practice in a way that is essentially *de facto* and *a posteriori*. In the final analysis, the credentials of science derive from strictly practical rather than from purely theoretical considerations.

The occurrences of nature (though not, to be sure, our *description* of them) do not impose some unique explanatory framework for their rationalization. The facts can be accommodated to various diverse modes of explanatory rationalization. The result is a choice between a plurality of diverse alternatives. This is the basis for an essentially "rationalistic" approach: We do not extract an explanatory framework from the facts—we bring one to them. A creative intellectual act thus lies at the base of explanation.

But although "the facts" do not *dictate* the outcome of the choice of some one explanatory framework they do *restrict* it. The selection of an explanatory framework is not an arbitrary act. Theory, here as everywhere, will stub its toes against the hard bedrock of brute fact. The choice thus becomes limited: events, and the regularities that obtain among them, are what they are, and our conceptualization, classification, and explanation of them cannot proceed to ignore this, or rather, can do so only at the peril of failure. And failure at this level is governed not by theoretical considerations relating to the abstract reasoning of explanation, but by pragmatic considerations of effectiveness or ineffectiveness on the side of practical issues: pre-eminently *prediction* of the future and *control* over nature.

The choice of a theoretical framework of explanation is thus a limited or circumscribed choice. It is a *genuine* choice because there are alternatives; but it is a *limited* choice because the range of viable alternatives is restricted. The situation is analogous to an underdetermined problem in mathematics of the type: Given two points, draw an equilateral figure of fewer than six sides with these points as vertices. Such examples of problems admitting a plurality of admissible solutions are the paradigm for a situation of limited choice.

But this plurality of alternatives is not the whole story, for the adoption of some one of these solutions is in this present case not a matter of haphazard. The choice at issue is not only limited, it is guided — principally by the criteria of pragmatic success in the areas of prediction and control.

This line of thought indicates how the findings of science can be secured against the assault of traditional philosophic scepticism. The sceptic's argument goes as follows: "The rational man must, of course, have a basis for his beliefs and opinions. Thus, asked *why* he accepts some accepted belief or opinion, he will cite one (or more) others that support it. But we can now ask him why he accepts these in turn, and this process can be continued as long as one likes. As a result, we will either move in a circle—and so ultimately provide no justification at all—or become involved in an infinite regress, supporting the elephant on the back of a turtle on the back of an alligator, etc. The only way to terminate the regress is by a dogmatic acceptance, somewhere along the line, of an *ultimate* belief that is used to justify others but is not itself justified. But any such unjustified acceptance is by its very nature arbitrary and irrational. Thus how can there ever be a secure standard of rational belief. But how can we do this save by appealing to another standard—and then again onwards until some ultimately unjustified standard is irrationally accepted, so that the adoption of any ultimate standard is ultimately arbitrary?" So reasons the philosophic sceptic.

Our answer to this line of scepticism lies in recognizing that the things one rationally accepts are not of a piece. Specifically, it is necessary to give a careful heed to the distinction between *theses* on the one hand and *methods* on the other. It is indeed ultimately unsatisfactory to justify theses in terms of further theses in terms of further theses, etc. But reflection on the rational justification of the scientific enterprise shows that this is not at all the issue here. Rather, we justify our acceptance of certain theses because (ultimately) they are validated by the employment of a certain method, the scientific method, thus breaking outside the cycle of justifying thesis by thesis by the fact that a thesis can be justified by application of a method. And we justify the adoption of this method in terms of certain *practical* criteria: success in prediction and efficacy in control. Our dialectic of justification thus breaks out of the restrictive confines of the sceptic's circle, and does so without relapse into a dogmatism of unjustified ultimates. We justify the acceptance of theses by reference to the method from which they derive, and we justify the method in terms of the classical pragmatic criterion. (With respect to *methodology*, at any rate, the pragmatists were surely right— there is certainly no better way of justifying a *method* than by establishing that "it works" with respect to the specific tasks held in view.) Our approach thus counters philosophical scepticism by a complex, two-stage maneuver, combining the methodological justification of scientific theses with a pragmatic justification of scientific method.

The upshot of the discussion of these last two sections may thus be summarized as follows. The endeavor of explaining the occurrences of the natural universe can in theory proceed along different alternative lines. The scientific framework of explanation is but one among others. Moreover, the choice of this particular framework is not forced upon us by the facts; we do not extract it from the facts, we bring it to them. The superiority of the scientific approach is not something that can be demonstrated on an *a priori* basis, and indeed does not ultimately rest upon theoretical considerations at all. Relating this final theme of our discussion to an earlier one, we can say that while prediction and control cannot be viewed as exclusively constituting "the aims" of science, they do represent criterial factors by reference to which the superiority of the scientific framework of explanation can be established.

appendix I

ARE HISTORICAL EXPLANATIONS DIFFERENT?*

1. Introduction

The claim is frequently made that the mode of explanation to be found in history differs radically and fundamentally from the types of explanation found in the natural or social sciences. This difference is said to lie in the fact that history, unlike science, must always deal with "the description of a situation or state of affairs which is unique."[1] It is argued that the exclusive objects of historical understanding are unique, particular, concrete events. The historian, it is contended, is primarily concerned with describing and analyzing the *unique* features of his data, unlike the scientist who looks to the *generic*. This view, if correct, would have far-reaching implication for the possibilities of scientific explanation in the historical domain, since such explanation— proceeding as it does by subsumption arguments—could never cope with factors that are nowise generic because utterly unique.

*This appendix is a revised version of a paper written in collaboration with Professor Carey B. Joynt of Lehigh University and initially published under the title "The Problem of Uniqueness in History" in *History and Theory*, vol. 1 (1961), pp. 150–162. I am grateful to Professor Joynt and to the editor of *History and Theory* for permission to use this material here.

1. W. Dray, *Laws and Explanation in History* (London, 1957), p. 44.

It is the thesis of this appendix that such a claim for history (which, when properly understood, contains a large measure of truth) cannot, without severe qualifications, survive objections which can be brought against it. Practicing historians, it is true, sometimes try to defend the claim of history to uniqueness, but as a rule these efforts are swept aside by the critics as special pleading that does not come to grips with the substantive arguments which can be marshaled against them.

In the interests of clarity and truth it should be frankly recognized, to begin with, that in a significant sense every particular event whatsoever is unique:

> Every insignificant tick of my watch is a unique event, for no two ticks can be simultaneous with a given third event. With respect to uniqueness each tick is on a par with Lincoln's delivery of the Gettysburg address!... Every individual is unique by virtue of being a distinctive assemblage of characteristics not precisely duplicated in any other individual.[2]

That is to say, it would seem to be an elemental fact about the universe that all events whatsoever are unique. Every concrete natural occurrence is unqualifiedly unique, even the occurrence of a so-called "recurrent" phenomenon like a sunrise or of "repeatable" events like the melting of a lump of sugar in a teacup.

Events are all unique in actuality; they are to be rendered non-unique *in thought only*, by choosing to use them as examples of a conceptualized type or class. We refer alike to "an appearance of a comet" or "a seafight with sailing ships," and our use of such terms, when examined, suggests that the occurrences of natural history (in the sense of nonhuman history) do not differ as regards uniqueness from the events of human history. Whether an event is selected for treatment as a unique, concrete particular, or is treated as the nonunique exemplar of a class of events, is essentially a matter of human interest and perspective. Galileo, rolling

2. A. Grünbaum, "Causality and the Science of Human Behavior," *American Scientist*, vol. 40 (1952), pp. 665–676; reprinted in M. Mandelbaum, F. W. Gramlich, and A. R. Anderson (eds.), *Philosophic Problems* (New York, 1957).

a ball down an inclined plane, treated each roll as identical, for it served his purposes so to do, just as an historian speaking of the Black Death could, if he wished, treat each unique death as identical in its contribution to a class of events called "a plague." Like the scientist, the historian resolves the dilemma of uniqueness by the use of a large variety of classes in his discussions: "nations," "wars," "revolutions," "assassinations," "budgets," and the like. The list is endless. It cannot be maintained that the use of such classes sets history apart in that they fail to exhaust the unique structure of a particular person or event, for exactly the same holds true of all scientific classification. Only *some* of the features of a given particular are described in such a classification, and no set of generic classifications could conceivably exhaust the structure of the particular objects or events so described.

Bearing these general but fundamental comments in mind, let us now examine in more detail the claim that history deals, in some sense, with unique events. Three particular issues can be utilized to throw considerable light upon this central problem: the relation of history to the "historical" sciences, the role of generalizations in history, and the requirements of interpretation in history.

2. The Relation of History to the "Historical" Sciences

History, it is clear, has no monopoly on the study of the past. The biologist who describes the evolution of life, the sociologist or anthropologist who delves into the development of human organizations and institutions, the philologist who analyzes the growth and change of languages, the geologist who studies the development of our planet, and the physical cosmologist who investigates the evolution of the cosmos—all deal with essentially "historical" questions, that is to say, with the events and occurrences of the past. Many diverse areas of scientific inquiry have the past as the "target" of their researches. It is therefore natural and appropriate to ask: How does history (history proper, i.e., *human* history) differ from these other "historical" sciences?

In attempting a reply to this question, we must recognize that it is simply not enough to insist that history deals with the doings of men in the context of civilized society, however true this remark may be. The biomedical student of human ecology is also concerned with man and his social environment, and of course anthropologists and sociologists study the activities of men in the context of human institutions, past no less than present. Consequently, there can be no adequate grounds for maintaining that history is to be distinquished from the "historical" sciences on the basis of subject-matter considerations alone.

Nor can a warrant for the distinction be said to inhere in the methodology of research. For here also there is no hard-and-fast barrier of separation between history and the sciences. History conforms fully to the standard hypothetico-deductive paradigm of scientific inquiry, usually described in the following four steps.

1. Examination of the data
2. Formulation of an explanatory hypothesis
3. Analysis of the consequences of the hypothesis
4. Test of these consequences against additional data

Historical research follows just this pattern; the historian assembles his chronological data, frames an interpretative hypothesis to explain them, examines the consequences of this hypothesis, and seeks out the additional data by which the adequacy of this hypothesis can be tested. The universal characteristics of scientific procedure characterize the work of the historian also. And even if the specific form which this generic process takes in history were to differ from that which it assumes in other areas of science in certain points of detail, this would be irrelevant to the matter at issue.[3] Botany has less need for the algebraic theory of

3. We do not mean to deny that in certain fields of scientific inquiry students can take advantage of their ability to repeat experiments, to devise so-called

matrices than does quantum mechanics or the economics of production processes; however, this does not mean that such differences in mathematical requirements can be used as a basis for claiming that these fields can be delimited on methodological grounds.

It thus appears that the definitive characteristics of history are to be found neither in its subject matter nor in its methods. Does it follow that there is no essential difference between history and the "historical" sciences? Not at all! But to see the precise character of the distinction we must examine closely the relative role of datum and theory, of fact and law in various sciences.

Throughout the various sciences, including all of those sciences which we have here characterized as "historical," we find that the object of the science is the study of a certain range of basic "fact" with a view to the discovery of generalizations, ideally universal laws, which govern the range of phenomena constituting this factual domain. In consequence, particular facts have a strictly *instrumental* status for the sciences. The "facts" serve as data, as means to an end: the law. In the sciences, the particular events that comprise the facts studied play an indispensable but nonetheless strictly subordinate role: The focus of interest is the general law, and the particular fact is simply a means to this end.

In history, on the other hand, this means-end relationship is, in effect, reversed. Unlike the scientist, the historian's interest lies first and foremost in the particular facts of his domain. But of course he is not solely interested in historical facts and in describing them. For this is mere chronology, which may constitute the inevitable starting point for history, but is by no means to be confused with history itself. The historian is not simply interested in dating events and describing them but in *under-*

---

crucial experiments, and that this is a privilege denied to all the historical sciences. This issue is too intricate to be discussed adequately in this appendix. We would only remind our readers that (1) nonrepeatability of experiments does characterize certain of the natural sciences also, and (2) so-called "thought experiments" are available to historians even as they are to physicists.

standing them. And "understanding" calls for interpretation, classification, and assessment, which can only be attained by grasping the relationship of casual and conceptual interrelation among the chronological particulars.

Failure to recognize this essential point undermines Dray's argument that it is of the essence of historical explanation that it deals with a "continuous series" of happenings and that this model of a continuous series best describes the offerings of historians.[4] In fact, a continuous series of events is, taken by itself, simply chronology — no less and no more. The following example, in the form of a continuous time series, illustrates the point at issue:

> "At 10 A.M. Napoleon finished his breakfast. At 10:20 he took off his night attire, called for his riding clothes, and at 10:35 kissed his wife good-bye. At 11 A.M. he attended a meeting of his commanders, decided to begin the campaign in ten days, and returned to his quarters at 1 P.M. At 1:02 P.M. he scratched his head, patted his forelock, and at 1:03 P.M. he sat down reflectively before the large portrait of a former mistress."

It is just at this point that scientific generalizations and laws enter upon the scene. They provide the necessary means for understanding particular facts: They furnish the fundamental patterns of interrelationships that constitute the links through which the functional connections among particular events may be brought to view.

Now if this analysis is correct, it is clear that the historian simply inverts the means-end relationship between fact and theory that we find in science. For the historian *is* interested in generalizations, and *does* concern himself with them. But he does so not because generalizations constitute the aim and objective of his discipline, but because they help him to illuminate the particular facts with which he deals. History seeks to provide an *understanding of specific occurrences*, and has recourse to such laws and generalizations — largely borrowed from the sciences but also

4. Dray, *op. cit.*, pp. 66–72.

drawn from ordinary human experience — which can be of service in this enterprise. But here the role of generalizations is strictly instrumental: They provide aids toward understanding particular events. The scientist's means-end relation of facts to laws is thus inverted by the historian.

Correspondingly, the very way in which history concerns itself with the past is quite different from that of the "historical" sciences. The historian is interested in the particular facts regarding the past *for themselves*, and not in an instrumental role as data for laws. Indeed, unlike the researcher in "historical" science, the historian is not a *producer* of general laws but a *consumer* of them. His position vis-à-vis the sciences is essentially parasitic. The generalizations provided by anthropology, sociology, psychology, and other sciences are used by the historian in the interests of his mission of facilitating our understanding of the past.

On this analysis, the line of distinction between history and the "historical" sciences is not an obvious but a rather subtle matter. Thus both the sociologist and the historian can take interest in precisely the same series of phenomena, say, the assimilation of Greek learning into medieval Islam. And both may bring essentially the same apparatus to bear on the study of the facts. But for the sociologist this is a "case study" undertaken in the interests of a general characterization of the generic process of the cultural transplant of knowledge. The study serves solely as an "input," as data for a mind seeking to provide rules governing this range of phenomena in general. Now this line of approach will also interest the historian, and any general rules provided by the sociologist would be material that the historian will gladly put to use in this study of Islamic cultural history. But his interest in such generalizations is purely instrumental as a means for explicating what went on in the particular case under study. Generalizations he must have, if he wishes to reveal interpretative links among events and to surpass mere chronology. But his aim is a clarification of the past *per se*, and his purpose is to provide an understanding of the past for its own sake, not merely as an instrument in the search for laws. The difference, then, between history and the "historical" sciences resides neither in subject matter nor in method, but in the objectives of the research and the consequent perspective that is taken in looking at the past. History does not collect facts to establish laws; rather, it seeks to exploit laws to explain facts. In this lies part of history's claim to uniqueness.

To obtain a clearer view of this essentially distinctive and characteristic relationship of fact and law in history, it will be necessary to examine more closely the role played by generalizations in history.

3. The Role of Generalizations in History

The question of the role of generalizations in history bears intimately upon the problem of the uniqueness of historical events. Since generalizations must, in the nature of things, deal with *types* or *classes* of events, it follows that they can have pertinence to specific, particular events only insofar as these are typical and classifiable, i.e., just insofar as they are *not* unique. Thus, determination of the extent to which generalizations can play a legitimate and useful role in history offers the best means of pin-pointing those ways in which historical events can properly be said to be "unique."

Perhaps the most famous thesis regarding the role of generalization in history is the doctrine which holds that there is one single grand law governing the fate of nations, empires, civilizations, or cultures. On such a view, all of the really major and large-scale transactions of history fall inevitably into one and the same basic pattern. The principal and characteristic function of the historian is to discern and then articulate this supreme generalization which pervades the regularity inherent in the historical process, a procedure familiar to readers, in the present century, through the work of Spengler and Toynbee. On a view of this nature, uniqueness plays at best a very restricted role in history, being limited to points of detail, in virtue of the predominant status of the basic pattern of regularity.

In reaction against this view, various theorists have taken their position at the very opposite extreme, and have espoused the view that generalization has no place whatever in history. On this conception, the supreme fact of historical study is the total absence of generalizations from the historical domain. Pervasive uniqueness becomes the order of the day. The historian, it is argued, inevitably deals with nonrepeatable particulars. Generalization is not only unwarranted, it drastically impedes all understanding of the data of history. Because it does away with the very possibility of historical explanation—for which some sort of generalizations would inevitably be required—this view may be characterized as "historical nihilism."

Our own view lies squarely between these two contrasting positions. The plain fact is that no grand generalization of the "pattern of history" has ever been formulated which is both (1) sufficiently specific to be susceptible to a critical test against the data and (2) sufficiently adequate to survive such a test. Thus it is not, in our judgment, to such a thesis that one can look for an acceptable account of the place of generalization in history. On the other hand, it is demonstrable that failure of the grand generalization approach to history in no way justifies ruling generalizations out of the historical pale. To see why this is so requires a closer look at the role of laws and generalizations in history.

To begin with, it is clear that the historian must make use of the general laws of the sciences. He cannot perform his job heedless of the information provided by science regarding the behavior of his human materials. The teachings of human biology, of medicine, or of psychology can be ignored by the historian only at his own peril. The facts offered by these sciences regarding the mortality and morbidity of men, their physical needs for nourishment, sleep, and so on, their psychological make-up, and the like represent essentially unchanging constants in the functioning of the human materials with which history deals. Nor can the physical sciences which describe the behavior of man's environment, again by means of general laws, be ignored by the historian. History must ever be rewritten if only because the progress of science leads inescapably to a deepening in our understanding of historical events. And because the general laws of the sciences deal with the fundamental constancies of nature, and not with idiosyncratic particulars, their role in historical understanding is a primary locus for *nonuniqueness* in history.

Going beyond this, it is important to recognize that the general laws of the natural sciences do not constitute the only basis of generalization in history. These sciences give us the characterization of the physical, biological, and psychological boundary conditions within which man must inevitably operate. But there are also sets of boundary conditions that stem from man's *cultural* rather than his physical environment. These bring to the scene of historical explanation the general laws of the social sciences, which create further conditions of nonuniqueness.

Nor is this the whole story. We must now consider another mode of generalization that has pivotal importance in historical explanation, namely generalizations which represent not the strictly universal laws to which we are accustomed from the sciences, but *limited* generalizations. Such limited generalizations are rooted in *transitory regularities,* deriving from the existence of temporally restricted technological or institutional patterns. Such technologically rooted transitory regularities illustrate how limited generalizations can arise in a way that makes it possible for them to have the feature of hypothetical force essential to their explanatory use. This at least exhibits the feasibility of "historical laws" of a limited sort—operative within specific spatiotemporal limits.

A second major source of limited generalizations is constituted by the entire sphere of institutional practices. Social customs, legal and political institutions, economic organizations, and other institutional areas, all constitute sources of such limited generalizations. Thus a limited generalization can be based, for example, upon the U.S. practices of holding a presidential election every four years and a population census every ten. Here again we have regularities which are limited, in temporal (and of course geographic) scope to an era during which certain fundamental institutional practices are relatively constant. Such institutional patterns will of course be of immense value to the historian in providing an explanation for the relevant events. He will seek to discover such institutional regularities precisely because they afford him the means to an explanation of occurrences. Events within a limited period can be understood and explained in terms of the limited generalizations which capture the particular institutional structure of this era. And the existence of institutionalized patterns of regularity again makes for a limitation upon the extent to which "uniqueness" obtains with respect to the data with which the historian deals.[5]

It should be noted, however, that the utilization in historical explanations of limited generalizations based upon reference to temporary technological and institutional eras does not provide the basis for a fundamental separation between history and the natural sciences. Such reliance upon transitory regularities does not make a place for uniqueness in history in any absolute sense. After all, the past stages of biological and cosmological evolution are also nonrepetitive. And thus the "historical" departments of the natural sciences must also deal with nonrecurring eras. And in these domains, limited generalizations can be — and sometimes are — also formulated. But as a rule the scientist is concerned with such limited regularities only as a waystation en route to the universal laws which are the main focus of his interest, and therefore tends, by and large, to be relatively aloof to the peculiarly limited generalizations which

5. This argument was set forth in C. B. Joynt and N. Rescher, "On Explanation in History," *Mind*, vol. 68 (1959), pp. 383–388.

also could be formulated for his domain. But the historian, whose interest must focus upon the understanding of particular events and not the formulation of universal generalizations, has a much larger stake in limited generalizations.

The upshot of the present analysis of the role of generalizations in history can thus be summarized as follows. History must and does use generalizations: first as a consumer of scientific laws, secondly as a producer of limited generalizations formulated in the interests of its explanatory mission and its focus upon specific particulars. The use of all of these types of generalizations in history is indispensible for the historian's discharge of his explanatory mission. Explanation demands that he be able to spell out the linkages between events — linkages of causation, of influence, and so on — and this can only be done in the light of connecting generalizations. To see this more clearly, and to attain a fuller perspective of the place of interpretation in history, we turn now to an analysis of what is involved in historical interpretation.

4. The Requirements of Explanation Interpretation in History

The considerations so far brought forward do not, it appears, fully meet the hard core of the claim so clearly and forcefully stated by Dray that "historical events and conditions are often unique simply in the sense of being different from others *with which it would be natural to group them under a classification term*" so that the historian is "almost invariably concerned with [an event] as *different* from other members of its class."[6] Thus the thesis claiming a characteristic role for uniqueness in history is based in the final analysis upon the assertion that historical events are "unique" in the sense that *all* classifications used will fail at a critical juncture to illuminate adequately the event in

6. Dray, *op. cit.*, p. 47.

question, since the event differs in its essential characteristics from its associates in these classes.[7]

A history composed of strictly nonrepeatable occurrences would be mere chronicle. Any efforts to establish connections among events and to exhibit their mutual relevance and significance would be excluded by a strict application of Dray's dictum. But, of course, history deals not only with what has happened, but with the *explanation* of what has occurred. The moment one leaves the bare statement of brute fact and occurrence and turns to interpretation, stark and atomic uniqueness is necessarily left behind. The inescapable fact is that the mere *record* of what happened does not suffice to *explain* anything.[8] Interpretation and explanation forces the historian to become involved in the problem of causation, and hence to abandon uniqueness.

An examination of any satisfactory historical account of a series of events shows clearly the indispensable role played by categories, classifications, and generalizations (whether universal or limited in range). Such concepts will contribute in two main ways to the reconstruction of events by the historian.

In the first place, the concepts enable the historian to *select* important and relevant facts from the infinite number of facts which are available to him. It often occurs that the body of pertinent fact is overwhelming in scope, and that the task of the historian requires judicious selection. No practicing historian has stated this proposition with greater clarity than Lord Macaulay in his *Essay on History*:[9]

> Perfectly and absolutely true it [history] cannot be: for, to be perfectly and absolutely true, it ought to record *all* the slightest

7. Compare the discussion by N. Rescher, "On the Probability of Non-Recurring Events," appearing in H. Feigl and G. Maxwell (eds.), *Current Issues in the Philosophy of Science* (New York, 1961), pp. 228–241.

8. Cf. J. H. Randall, Jr., *Nature and Historical Experience* (New York, 1958), pp. 61–64.

9. *The Works of Lord Macaulay*, ed. by his sister Lady Trevelyan, vol. 5 (New York, 1897), pp. 129–130.

particulars of the slightest transactions—all the things done and all the words uttered during the time of which it treats. The omission of any circumstance, however insignificant, would be a defect. If history were written thus, the Bodleian library would not contain the occurrences of a week. What is told in the fullest and most accurate annals bears an infinitely small proportion to what is suppressed. The difference between the copious work of Clarendon and the account of the civil wars in the abridgment of Goldsmith vanishes when compared with the immense mass of facts respecting which both are equally silent. No picture, then, and no history, can present us with the whole truth: but those are the best pictures and the best histories which exhibit such parts of the truth as most nearly produce the effect of the whole. He who is deficient in the art of selection may, by showing nothing but the truth, produce all the effect of the grossest falsehood.

Without the guidance of generalizations and general concepts, the historian would be trapped and bogged down—drowned if you will—in the welter of concrete particulars. A general concept, such as that of a "General Staff" and its corporate role in military decisions, enables historians to organize their data, emphasizing some and putting aside a multitude of details which have little or no bearing upon the choices eventually made. The concept of "an alliance" in turn permits a writer to put into place a whole series of events directly relevant to the eventual course of events, and to erect a coherent structure upon it.

Secondly, such generic concepts have particular importance of the *explanation* of the events in question. Their use essentially determines the meaning and significance of the account which finally emerges from the historian's efforts. They are, in short, absolutely crucial to the attempt of the historian to interpret his material, to raise it above the level of narrative and chronology to the status of true history. By utilizing, in the example just cited, the concepts of alliance systems, and of the organizational role of the military in their impact upon civilian statesmen, the historian can demonstrate that France believed the Triple Entente was necessary to her survival as a Great Power, and that

Germany's policy decisions were heavily influenced by the general strategy of the Schlieffen plan. In this way, definite casual patterns are erected through the use of general concepts. Similarly, it can be demonstrated that the existing technology of the period played a vital part in the eventual decisions of the various powers to order general mobilization, or to issue an ultimatum to their opponents to cease such a procedure lest war follow. Here then we find limited generalizations serving to delineate the institutional boundary conditions under which men operated, enabling the historian to set up a general pattern into which particular events can be fitted and from which certain definite conclusions can justifiably be derived.

There can be little doubt that the use of categories, classes, and generalizations are absolutely essential to, and perform a vital function for, the adequate discharge of the historian's task. Indeed, taken together, they constitute the framework and structure of history, the setting in which the recital of particulars unfolds. They constitute the hard core of explanation and interpretation and, by their presence and the vital nature of the role they play, go a very long way indeed toward qualifying the claim that history deals solely or even primarily with the unique features of particular events.

5. History and the Problem of Prediction

In closing, some remarks about the predictive aspect of history are in order. To say without qualification that history cannot predict is plainly false. Historical predictions are actually a commonplace of modern life, and it is a matter of common knowledge that many important developments in human affairs can be foretold with great accuracy on the basis of an historical analysis of past trends (e.g., demographic facts regarding life expectancies, population densities of cities and countries, or even such cultural phenomena as the number of books to be published in a given country in a given subject-matter field). What is meant by the dictum that "history cannot predict" is that it cannot forecast those major critical developments on the world scene *in which we are most interested* and in which our curiosity is the keenest (in part precisely because they are so difficult to foretell). The situation is similar in the field of medicine, which can predict both the very near future (no change) and the very distant (we'll all be dead), but usually cannot answer the really "interesting" questions about our state of health a matter of months or even years ahead.

Now if it be conceded that history cannot predict the future in the sense we have specified, wherein does the value of historical understanding lie? If history cannot predict, is not its interest purely antiquarian because the study of historical explanation cannot now serve any practical need?

To ask the question in this way seems to me to insist upon looking right past what is the one valid pragmatic justification of historical studies. This, we maintain, has nothing to do with knowledge about the future. It is not the task of history to furnish some crystal ball in which the shape of things to come may be discerned. The value of history in relation to the future lies in its providing not knowledge but wisdom. Historical understanding does not tell us what will happen. Rather, by providing us with a background for understanding the behavior of human individuals, societies, and institutions in their reactions to challenges and opportunities under the most diverse circumstances, history places us in a better position to react with intelligence, balanced perspective, and good sense to whatever does happen, no matter what this may be.

appendix II

ON THE EPISTEMOLOGY OF THE INEXACT SCIENCES *

"The laws of economics are to be compared with the laws of the tides, rather than with the simple and exact law of gravitation. For the actions of men are so various and uncertain, that the best statement of tendencies, which we can make in a science of human conduct, must needs be inexact and faulty. This might be urged as a reason against making any statements at all on the subject; but that would be almost to abandon life. Life is human conduct, and the thoughts and emotions that grow up around it. By the fundamental impulses of our nature we all — high and low, learned and unlearned — are in our several degrees constantly striving to understand the courses of human action, and to shape them for our purposes, whether selfish or unselfish, whether noble or ignoble. And since we *must* form to ourselves some notions of the tendencies of human action, our choice is between forming those notions carelessly and forming them carefully. The harder the task, the greater the need for steady patient inquiry; for turning to account the experience, that has been reaped by the more advanced physical sciences; and for framing as best we can well thought-out estimates, or provisional laws, of the tendencies of human action."

— ALFRED MARSHALL, *Principles of Economics*, 1892

*This appendix is a revised version of a paper that was written in collaboration with Olaf Helmer in 1956 and which appeared under the same title as an article in *Management Science*, vol. 6 (1959), pp. 25–52. I am grateful to Dr. Helmer and to the editor of *Management Science* for permission to use this material here.

1. The Mythology of Exactness

It is a fiction of long standing that there are two classes of sciences, the exact and the inexact, and that the social sciences by and large are members of the second class—unless and until, like experimental psychology or some parts of economics, they "mature" to the point where admission to the first class may be granted.

This widely prevalent attitude seems to us fundamentally mistaken; for it finds a difference in principle where there is only one of degree, and it imputes to the so-called exact sciences a procedural rigor that is rarely present in fact. For the sake of a fuller discussion of these points, let us clarify at the very outset the terms "science," "exact science," and "inexact science," as they are intended here.

For an enterprise to be characterized as *scientific* it must have as its purpose the explanation and prediction of phenomena within its subject-matter domain and it must provide such explanation and prediction in a reasoned, and therefore intersubjective, fashion. We speak of an "exact science" if this reasoning process is formalized in the sense that the terms used are exactly defined and reasoning takes place by formal logico-mathematical derivation of the hypothesis (the statement of the fact to be explained or predicted) from the evidence (the body of knowledge accepted by virtue of being highly confirmed by observation). That an exact science frequently uses mathematical notation and concerns itself with attributes that lend themselves to numerical measurement we regard as incidental rather than defining characteristics. The same point applies to the precision, or exactness, of the predictions of which the science may be capable. While precise predictions are indeed to be preferred to vague ones, a discipline that provides predictions of a less precise character, but makes them correctly and in a systematic and reasoned way, must be classified as a science.

In an inexact science, conversely, reasoning is informal; in particular, some of the terminology may, without actually impeding communication, exhibit some inherent vagueness, and reasoning may at least in part rely on reference to intuitively perceived facts or implications. Again, an inexact science rarely uses mathematical notation or employs attributes capable of exact measurement, and as a rule does not make its predictions with great precision and exactitude.

Using the terms as elucidated here — and we believe that this corresponds closely to accepted usage — purely descriptive surveys or summaries, such as the part of history that is mere chronology or, say, purely descriptive botany or geography, are not called "sciences." History proper, on the other hand, which seeks to explain historical transactions and to establish historical judgments having some degree of generality, is a science; it is in fact largely coincident with political science, except that its practitioners focus their interest on the past, while the political scientists' main concern is the present and the future.

As for *exactness*, this qualification, far from being attributable to all of the so-called natural sciences, applies only to a small section of them, in particular to certain subfields of physics, in some of which exactness has even been put to the ultimate test of formal axiomatization. In other branches of physics, such as parts of aerodynamics and of the physics of extreme temperatures, exact procedures are still intermingled with unformalized expertise. Indeed the latter becomes more dominant as we move away from the precise and usually highly abstract core of an exact discipline and toward its applications to the complexities of the real world. Both architecture and medicine are cases in point. Aside from the respective activities of building structures and healing people, both have a theoretical content — that is, they are explanatory and predictive ("This bridge will not collapse, or has not collapsed, because . . . ." "This patient will exhibit, or has exhibited, such and such symptoms because . . .").

They must therefore properly be called "sciences," but they are largely inexact since in actual operation they rely extensively on informal reasoning processes.

If in addition to these examples we remember the essentially intermediate status of such fields as economics and psychology, both of which show abundant evidence of exact derivations as well as reliance on intuitive judgment (exhibiting intermittent use of mathematical symbolism and of measurable attributes and an occasional ability to predict with precision), it should be obvious that there is at present no clear-cut dichotomy between exact and inexact sciences, and, in particular, that inexactness is not an attribute of only the social sciences.

However, leaving aside their present comparative status, it still might be possible to hold the view that there exists an epistemological difference *in principle* between the social sciences on the one hand and the natural or physical sciences on the other, in the sense that the latter, though not necessarily quite exact as yet, will gradually achieve ultimate exactness, while the former, owing to the inherent intractability of their subject matter and the imperfection in principle of their observational data, must of necessity remain inexact. Such a view would be based on a false premiss, namely, a wholly misguided application of the exactness-versus-inexactness distinction. Indeed, the artificial discrimination between the physical sciences with their (at least in principle) precise terms, exact derivations, and reliable predictions, as opposed to the social sciences with their vague terms, intuitive insights, and virtual unpredictability has retarded the development of the latter immeasurably.

The reason for this defeatist point of view regarding the social sciences may be traceable to a basic misunderstanding of the nature of scientific endeavor. What matters is not whether or to what extent inexactitudes in procedures and predictive capability can eventually be removed (with regard to predictive precision, the social sciences may perhaps always fall short of the physical sciences); rather it is *objectivity*—that is, the intersubjectivity

of findings independent of any one person's intuitive judgment —
that distinguishes science from intuitive guesswork, however
brilliant. This has nothing to do with the intuitive spark that may
be the origin of a new discovery; pure mathematics, whose formal
exactness is beyond question, needs that as much as any science.
But once a new fact or a new idea has been conjectured, no matter
on how intuitive a foundation, it must be capable of objective
test and confirmation by anyone. And it is this crucial standard
of scientific objectivity, rather than any purported criterion of
exactitude, to which the social sciences conform only imperfectly.

In rejecting precision of form or method as well as degree
of predictability as basic discriminants between the social and
the physical sciences, it thus remains to be seen whether there
might not in fact be a fundamental epistemological difference
between them with regard to their ability to live up to the same
rigorous standard of objectivity. Our belief is that there is essentially no such difference, in other words, that the social sciences
cannot be separated from the physical sciences on methodological grounds. We hope to convince the reader of the validity
of our position by offering, in what follows, at least some indications as to how the foundations for a uniform epistemology
of all of the inexact sciences might be laid — be they social
sciences or "as yet" inexact physical sciences.

2. Plan of Procedure

Our goal is more modest than that of presenting a comprehensive epistemology of the inexact sciences. We merely wish to
outline an epistemological attitude toward them that we would
like to see adopted more widely. Since epistemology is concerned
with the role of evidence in the attainment of scientific laws and
with the scientific procedures implied by that role, we need to
re-examine the status of such things as laws, evidence, confirmation, prediction, and explanation, with special reference to the
case of inexact sciences.

We shall begin with a brief look at historical laws, as typical examples of the explanatory generalizations encountered in the social sciences. They will typically be found to involve escape clauses of the "other things being equal" kind, which makes them different from the neat laws of the exact physical sciences; and the possibility of their application despite this defect will have to be discussed. In this context we shall examine the role of laws in prediction and in explanation and try to clarify the similarities and differences in the structure of prediction and explanation.

An analysis of this kind necessarily involves the prominent use of the notion of probability. In order to avoid misunderstandings, we shall sketch briefly our own views regarding this somewhat controversial concept. We prefer to distinguish — in addition to familiar statistical concept of probability based on observed frequencies — an *objective* and a *subjective* notion of probability, the former called "degree of confirmation," the latter, "personal probability"; the two will be seen to be linked via the concepts of "rational person" and "fair bets."

We shall then use these considerations regarding probability to clarify some difficulties arising in the sue of evidence for predictive purposes, and consider ways in which these difficulties may be surmounted by supplementing the objective evidence by judicious reliance on apparently subjective means (personal probabilities). This deviation from the standard procedures of the exact sciences will necessitate examining the possible justification of reliance on expert judgment and a search for improved and systematic methods for its use. Proposals for obtaining a consensus among a group of experts will be examined, as well as controlled opinion-feedback techniques that can aid their performance as a group. We next take up explicitly the question of whether, and if so to what extent, the standard procedures of the exact sciences have to be further augmented to meet the needs of the inexact sciences. Here we shall, in particular, discuss the concept of simulation models and their perhaps most important subcase, operational gaming models, together with the use of experts in conducting pseudo-experiments within such models. Finally, some suggestions will be made as to how the application of relatively unorthodox methods (such as the use of experts and gaming), tried successfully in other inexact sciences, might aid the applied social sciences, especially in exercising their predictive function with regard to decision-making processes.

3. Historical Laws

Let us, then, first take a brief look at historical science in order to obtain some illustrative examples of the form of laws in the social sciences and of the function they perform. A *historical law* may be regarded as a well-confirmed statement concerning the actions of an organized group of men under certain restrictive conditions (such group actions being intended to include those of systems composed conjointly of men and nonhuman instrumentalities under their physical control). Examples of such laws are "A census takes place in the United States in every decade year"; "Heretics were persecuted in 17th-century Spain"; and "In the sea fights of sailing vessels in the period 1653–1803, large formations were too cumbersome for effectual control as single units." Such statements share three features of epistemological importance and interest: they are *lawful, spatio-temporally restricted*, and *loose*. These features require elaboration.

To consider lawfulness, let us take for example the statement about the cumbersomeness of large sailing fleets in sea fights. On first view, this statement might seem to be a mere descriptive list of characteristics of certain particular engagements: a shorthand version of a long conjunction of statements about large-scale engagements during the century and a half from Texel (1653) to Trafalgar (1803). This view is incorrect, however, because the statement in question is more than an assertion regarding characteristics of certain actual engagements. Unlike a mere description, it can serve to explain developments in cases to which it makes no reference. Furthermore, the statement has counterfactual force. It asserts that in literally any large-scale fleet action fought under the conditions in question (with sailing vessels of certain types, having particular modes of armament and using contemporaneous communications methods), effectual control of a great battle line is hopeless. It is claimed, for example, that had Villeneuve issued from Cadiz some days earlier or later he would all the same have encountered difficulty in the management of the great allied battle fleet of over thirty sail of the line, and Nelson's strategem of dividing his force into two virtually independent units acting under prearranged plans would have facilitated effective management equally well as at Trafalgar.

The statement in question is thus no mere descriptive summary of particular events; it functions on the more general plane of lawful statements, specifically in that it can serve as a basis for explanation and that it can exert counterfactual force. To be sure, the individual descriptive statements that are known and relevant do provide a part of the appropriate evidence for the historical generalization. But the content of the statement itself lies beyond the sphere of mere description and, in taking this wider role, historical laws become marked as genuinely lawful statements.

Nevertheless, such historical generalizations are not unrestricted or universal in the manner in which the laws of the physical sciences are; they are not valid for all times and places. A historical law is limited, either tacitly or expressly, to applicability within specific geographic and temporal bounds. Usually historical laws are formulated by explicit use of proper names: names of places, of groups of persons, of periods of time, of customs or institutions, of systems of technology, of cultures, or the like. The restriction of application in such cases is overt in the formulation of the law. Sometimes, however, historical laws are formulated as unconditionally universal statements. But in such cases the statement properly interpreted takes on a conditional form of such a kind that its applicability is *de facto* tightly restricted. If sailing ships and contemporary naval technology and ordnance *were* reinstated, the tactical principles developed from the time of Tromp and de Ruyter to that of Rodney and Nelson would prove valid guidance. But the applicability of these tactical principles depends on the fulfillment of conditions that cannot reasonably be expected to recur. (This is not true of certain experimental situations in physics — such as the Michelson-Morley experiment — which merely are rarely repeated.) Thus although explicit use of proper names might be avoided in the formulation of a historical law, the statement is restricted nonetheless by conditions effecting in an oblique manner the same delimitation of its applicability to specific geographic and temporal bounds.

Finally, an important characteristic of historical laws lies in their being "loose." It has been said already that historical laws are (explicitly or obliquely) conditional in their logical form. However, the nature of these conditions is such that they can often not be spelled out fully and completely. For instance, the statement about sailing-fleet tactics has (among others) an implicit or tacit condition relating to the state of naval ordnance in the 18th century. In elaborating such conditions, the historian delineates what is typical of the place and period. The full implications of such reference may be vast and inexhaustible; in our example, ordnance ramifies *via* metalworking technology into metallurgy, mining, and so on. Thus the conditions that are operative in the formulation of a historical law may only be indicated in a general way and are not necessarily (indeed in most cases they cannot be expected to be) exhaustively articulated. This characteristic of such laws is here designated as *looseness*.

It is this looseness of its laws that typifies history as an inexact science in the sense in which we have used the term: in a domain where laws are not fully and precisely articulated there exists a limit to exactitude in terminology and reasoning. In such a sphere, mathematical precision must not be expected. To say this implies no pejorative intent whatever, for the looseness of historical laws is clearly recognized as being due, not to slipshod formulation of otherwise precise facts, but to the fundamental complexities inherent in the conceptual apparatus of the domain.

A consequence of the looseness of historical laws is that they are not universal, but merely quasi-general, in that they admit exceptions. Since the conditions delimiting the area of application of the law are often not exhaustively articulated, a supposed violation of the law may be explicable by showing that a legitimate (but as yet unformulated) precondition of the law's applicability is not fulfilled in the case under consideration. The laws may be taken to contain a tacit caveat of the "usually" or "other things being equal" type. A historical law is thus not strictly universal in that it must be taken as applicable to all cases falling within the scope of its explicitly formulated conditions; rather it may be thought to formulate relationships that obtain generally, or better, *as a rule*.[1]

Such a "law" we will term a *quasi-law*. In order for a quasi-law to be valid, it is not necessary that there be no apparent exceptions; it is only necessary that if an apparent exception should occur, an adequate explanation be forthcoming, an explanation demonstrating the exceptional characteristic of the case at hand by establishing the violation of an appropriate (if hitherto unformulated) condition of the law's applicability.[2]

1. This point has been made by various writers on historical method. Charles Frankel, for example, puts it as follows in his lucid article on "Explanation and Interpretation in History": "It is frequently misleading to take statements such as 'Power corrupts, and absolute power corrupts absolutely,' when historians use them, as attempts to give an exact statement of a universal law . . . . But such remarks may be taken as statements of strategy, rules to which it is best to conform in the absence of very strong countervailing considerations." (*Philosophy of Science*, vol. 24 [1957], p. 142.)

2. In his book *The Analysis of Matter* (London, 1927), Bertrand Russell writes: "Our prescientific general beliefs are hardly ever without exceptions; in science, a law with exceptions can only be tolerated as a makeshift" (p. 191).

For example, the historical law that in the pre-revolutionary French navy only persons of noble birth were commissioned is not without apparent exceptions, since in particular the regulation was waived in the case of the great Jean Bart, son of a humble fisherman, who attained great distinction in the naval service. We may legitimately speak here of an *apparent* exception; for instead of abandoning this universal law in view of the cited counterexample, it is more expedient to maintain the law but to interpret it as being endowed with certain tacit amendments which, fully spelled out, would cause the law to read somewhat as follows. "In the pre-revolutionary French navy, as a rule, only persons of noble birth were commissioned — that is, unless the regulation was explicitly waived or an oversight or fraud occurred or some other similarly exceptional condition obtained." While it may be objected that such a formulation is vague — and indeed it is — it cannot be said that the law is now so loose as to be vacuous; for the intuitive intent is clear, and its looseness is far from permitting the law's retention in the face of just *any* counterexample.[3] Specifically, if a reliable source brings to light one counterinstance for which there is no tenable explanation whatsoever to give it exempt status, a historian may still wish to retain the law in the definite expectation that some such explanation eventually will be forthcoming; but should he be confronted with a succession or series of unexplained exceptions to the law, he would no doubt soon feel compelled to abandon the law itself.

We thus have the indisputable fact that in a generally loose context, that of history being typical of the inexact sciences, it would be hopeless to try to erect a theoretical structure that is logically, perhaps even esthetically, on a plane with our idealistic

---

But this is true only in the physical sciences, if at all. A far juster view is that enunciated in Alfred Marshall's classic *Principles of Economics* (1892), cited in our motto.

3. Michael Scriven, "Truisms as the Grounds for Historical Explanation" in P. Gardiner (ed.), *Theories of History* (New York, 1959), pp. 443–475, speaks of historical generalizations as having a "selective immunity to counterexamples."

image of an exact theory. Yet, if we consider the situation not from the standpoint of the wishful dreamer of neat and tidy theory construction, but from that of the pragmatist in pursuit of a better understanding of the world through reasoned methods of explanation and prediction, then we have good reason to take heart at the sight of even quasi-laws; and we should realize that the seemingly thin line between vagueness and vacuity is solid enough to permit the distinction of fact from fiction reasonably well in practical applications.

4. Quasi-Laws in the Physical Sciences

We have chosen to illustrate the nature of limited generalizations (quasi-laws) by means of the graphic example of historical laws. Use of this example from a social-science context must not, however, be construed as implying that quasi-laws do not occur in the natural, indeed even in the physical, sciences. In many parts of modern physics, formalized theories based wholly on universal principles are (at least presently) unavailable, and use of limited generalizations is commonplace, particularly in applied physics and engineering.

Writers on the methodology of the physical sciences often bear in mind a somewhat antiquated and much idealized image of physics as a very complete and thoroughly exact discipline in which it is never necessary to rely on limited generalizations or expert opinion. But physical science today is very far from meeting this ideal. Indeed the laws of some branches of the social sciences are no less general than those of various branches of physics, such as the theory of turbulence phenomena, high-velocity aerodynamics, or extreme temperatures. Throughout applied physics in particular, when we move (say, in engineering applications) from the realm of idealized abstraction ("ideal" gases, "homogeneous" media) to the complexities of the real world, reliance on generalizations that are, in effect, quasi-laws becomes pronounced. (Engineering practice in general is based on "rules of thumb" to an extent undreamed of in current theories of scientific method.)

Thus no warrant whatever exists for using the presence of quasi-laws in the social sciences as validating a methodological separation between them and the physical sciences. A realistic assessment of physical-science methods shows that quasi-laws are operative here too. and importantly so.

With this in mind, let us now turn to a closer examination of the role played by laws — or quasi-laws — in explanation and prediction.

5.   Explanation and Prediction

A somewhat simplified characterization of scientific explana-tion — but one that nonetheless has a wide range of applicability, particularly in the physical sciences — is that explanation consists in the *logical derivation* of the statement to be explained from a complex of factual statements and well-established general laws. One would, for example, explain the freezing of a lake by adducing (1) the fact that the temperature fell below 32°F and (2) the law that water freezes at 32°F. These statements, taken together, yield the statement to be explained deductively.[4]

This deductive model of explanation, although adequate for many important types of explanations encountered in the sciences, cannot without serious emendation be accepted as applying to all explanations. For one thing there are proba-bilistic explanations, which can be based on statistical (rather than strictly universal) laws. ("I did not win the Irish Sweep-stakes because the chances were overwhelmingly against my doing so.") And then there are the quasi-laws occurring in the

4. For a full discussion of this matter, see C. G. Hempel and P. Oppenheim, "Studies in the Logic of Explanation," *Philosophy of Science*, vol. 15 (1948), pp. 135–175.

inexact sciences that because of their escape clauses cannot serve as the basis of strict *derivation* and yet can carry explanatory force. For example, the quasi-law cited earlier surely explains—in the accepted sense of the word—why the French fleet that supported Washington's Yorktown campaign was commanded by a nobleman (namely, the Comte de Grasse).

The uncertainty of conclusions based on quasi-laws is not due to the same reason as that of conclusions based on statistical laws. A statistical law asserts the presence of a characteristic in a certain (presumably high) percentage of cases; a quasi-law asserts it in all cases for which an exceptional status (in an ill-defined but clearly understood sense) cannot be claimed. We note for the moment, however, that the schema of explanation when either type of nonuniversal law is involved in the same, and in fact identical with what it would be were the law universal; and an explanation is regarded as satisfactory if, while short of logically *entailing* the hypothesis, it succeeds in making the statement to be explained highly *credible* in the sense of providing convincing evidence for it. We shall presently return to a discussion of the concept of evidence.

With regard to prediction as opposed to explanation, analyses of scientific reasoning often emphasize the similarities between the two, holding that they are identical from a logical standpoint, inasmuch as each is an instance of the use of evidence to establish a hypothesis, and the major point of difference between them is held to be that the hypothesis of a prediction or of an explanation concerns respectively the future or the past. This view, however, does not do justice to several differences between prediction and explanation that are of particular importance.

First of all, there are such things as *unreasoned* predictions—predictions made without any articulation of justifying argument. The validation of such predictions lies not in their being supported by plausible arguments, but may, for example, reside in proving sound *ex post facto* through a record of successes on the part of the predictor or predicting mechanism.

It is clear that such predictions have no analogue in explanations; only reasoned predictions, based on the application of established theoretical principles, are akin to explanations. However, even here there is an important point of difference.

An explanation should ideally *establish* its conclusion, showing that there is a strong warrant why the fact to be explained — rather than some possible alternative — obtains. On the other hand, the conclusion of a (reasoned) prediction need not be well established in this sense; it suffices that it be rendered *more tenable than comparable alternatives*. Here then is an important distinction in logical strength between explanations and predictions: A satisfactory explanation, though it need not logically rule out alternatives altogether, should as a minimum establish its hypothesis as *more credible than its negation*. Of a prediction, on the other hand, we need to require only that it establish its hypothesis as *more credible than any comparable alternative*. Of course predictions may, as in astronomy, be as firmly based in fact and as tightly articulated in reasoning as any explanation. But this is not a general requirement to which predictions *must* conform. A doctor's prognosis, for example, does not have astronomical certitude, yet practical considerations render it immensely useful as a guide in our conduct because it is far superior to reliance on guess work or on pure chance alone as a basis for decision-making.[5]

Generally speaking, in any field in which our ability to forecast

---

5. On the contrast between prediction and explanation see further I. Scheffler, "Explanation, Prediction, and Abstraction," *British Journal for the Philosophy of Science*, vol. 7 (1957) pp. 293–309, and N. Rescher, "On Prediction and Explanation," *British Journal for the Philosophy of Science*, vol. 8 (1958) pp. 281–290. The (somewhat oversimplified) position we have assumed in these paragraphs can be put more sharply — in terms of the distinctions developed in Part II above — as follows: that in the context of the inexact sciences a very weak probabilistic basis (of the P_w-type) might well be regarded as adequate for predictive but certainly not for explanatory purposes, whereas a strong probabilistic basis (of the P_s-type) is at the very least requisite to an achievement of scientific adequacy.

with precision is very limited, our actions of necessity are guided by only slight differences in the probability that we attach to possible future alternative states of the world; consequently we must base predictions on far weaker evidence than satisfactory explanations. This is especially true of a science such as history, or rather its predictive counterpart — political science. Here, in the absence of powerful theoretic delimitations that narrow down the immense variety of future possibilities to some manageable handful, the *a priori* likelihood of any particular state of affairs is insignificant, and we can thus tolerate considerable weakness in our predictive tools without rendering them useless. Consider, for example, the quasi-law that in a U.S. off-year election the opposition party is likely to gain, This is certainly not a general law, nor is it a mere summary of observed statistics. It has implicit qualifications of the "other things being equal" type, but it does claim to characterize the course of events "as a rule" and it generates an expectation of the explainability of deviations. On this basis, a historical (or political) law of this kind can provide a valid, though limited, foundation for sound predictions.

The epistemological asymmetry between explanation and prediction has not, it would seem, been adequately recognized and taken into account in discussions of scientific method. For one thing, such recognition would lead to a better understanding of the promise of possibly unorthodox items of methodological equipment, such as quasi-laws, for the purposes of prediction in the inexact sciences. But more generally it would open the way to explicit consideration of a *specific methodology of prediction* — a matter that seems to have been neglected to date by the philosophers of science. As long as one believes that explanation and prediction are strict methodological counterparts, it is reasonable to press further with solely the explanatory problems of a discipline, in the expectation that only the tools thus forged will then be usable for predictive purposes. But once this belief is rejected, the problem of a specifically predictive method arises, and it becomes pertinent to investigate the possibilities of predictive procedures autonomous of those used for explanation.

Before discussing such possiblities in greater detail, it is imperative, to avoid various misunderstandings, that we clarify briefly the meaning of probability and of some associated concepts.

6. Probability

From the viewpoint of the philosophy of science, the theory of probability occupies a peculiar position. To the extent that it deals with relations among propositions it is part of semantics and thus of pure logic. To the extent that it deals with credibility, rational beliefs, and personal expectations, it is part of empirical pragmatics and thus a social science. (The view, not held by us, that probability theory properly belongs entirely in the second field rather than the first is sometimes referred to as *psychologism*.[6]) Even for the logical part of the theory, the foundations are not yet established very firmly, and only in applications to the very simplest forms of idealized languages has real progress been made to date.[7] For this reason, some vagueness must still be accepted even in discussing the purely logical aspects of probability — that is, unless we are content to confine ourselves to the simplest case just mentioned, which we are not because the linguistic demands of the inexact sciences transcend these limits of simplicity even more frequently than do those of the exact sciences.

It is convenient to distinguish three probability concepts, namely, *relative frequency, degree of confirmation,* and *personal*

6. An incisive critique of psychologism is given in Chap. 2 of R. Carnap's book *Logical Foundations of Probability* (Chicago, 1950; second ed., 1960).

7. See again R. Carnap's massive study of the *Logical Foundations of Probability*, and various studies cited by him in the extensive bibliography, in particular those of Helmer, Hempel, and Oppenheim.

(or *subjective*) *probability*. Of these, the first is an objective, empirically ascertainable property of classes of physical objects or physical events; the second is also purely objective, namely, a logical relation between sentences; the third is a measure of a person's confidence that some given statement is true and is thus an essentially subjective matter. Let us briefly consider each of these three probability concepts.

Relative Frequency   Relative frequency requires the statement of a reference class (of objects or events), also called the "population." If the class is finite, the relative frequency is simply the ratio of the number of elements having some property or trait, divided by the total number of elements in the class. Thus we speak of the relative frequency of males in the present U.S. population, or of rainy days in Los Angeles in the first half of this century.[8] Sometimes the notion of relative frequency is extended to classes of either indefinite or infinite size. For example, we may speak of the relative frequency of male births in the United States over an extended period, without precisely specifying that period; or we may speak of the relative frequency "in the long run" of "heads" in tosses with a particular coin, where the sequence of tosses is of indefinite length and may even be idealized into an infinite sequence (in which case the "relative frequency" is the limit of the relative frequencies of the finite subsequences). In a situation like this, it is even customary to ascribe this probability, that is, the relative frequency of "heads" in the long run, as a property of the coin itself (in particular, a "fair coin" is one for which this probability is one half). But it is best to interpret such a statement merely as a paraphrase for the longer statement that in a long sequence of possible tosses with this coin (but not so long as to alter the physical characteristics of the coin) the relative frequency of "heads" will be one half.

8. It is a technical refinement, into which we need not enter here, that in applications it is common to use, instead of the relative frequency proper, some statistical *estimate* thereof.

Degree of Confirmation The degree of confirmation is a logical relation between two sentences, the hypothesis H and the evidence E. The degree of confirmation of H on the basis of E is intended to be a measure of the credibility rationally imparted to the truth of H by the assumed truth of E. Precise definitions have thus far been suggested only for the one-place predicate calculus. In the simplest case, where E has the form of a statistical record of n observations, to the effect that exactly m out of n objects examined had a property P, and where the hypothesis H ascribes this property P to an as yet unexamined object, the degree of confirmation of H on the basis of E, or $dc(H,E)$, is defined to be either the observed relative frequency m/n, or else a quantity very close to it (and having the same limit as n becomes large), which may differ somewhat from m/n because of technical requirements of elegance of the formalism. It is irrelevant for our present purposes which particular definition we adopt, but to fix the idea let us assume simply that in the above case $dc(H,E) = m/n$.

If E does not have the simple form of a statistic or H does not just affirm another like instance, then some plausible extension of the definition of "dc" is required; this may lead to cases where no single number can reasonably be specified but where the evidence merely warrants a narrowing down of the probability of H to several possible numbers or an interval of numbers. For instance, if H is the hypothesis that a certain Irish plumber will vote Democratic in the next presidential election, and evidence E amounts solely to saying that 70 per cent of the Irish vote Democratic and 20 per cent of the plumbers do, then all that can reasonably be asserted is that the required probability lies somewhere between .2 and .7.

Ambiguities of this kind can, of course, be removed by fiat (and in fact this has been the path followed in the formalisms proposed to date by Carnap). That is to say, one can transfer the ambiguity from the object language to the meta-language, by stating the matter as follows. There are several ways in which "dc" can be defined, but under each particular definition the degree of confirmation is a single-valued function.

No matter which of these alternatives is chosen, the situation can still be resolved, as long as we are dealing with one-place predicates only. When we move into a subject matter where adequate discourse requires multiplace predicates of several logical levels, no formal proposals for an extended definition of "dc" are as yet available, and we have to rely largely on trained intuition as to how a numerical measure of the "credibility rationally imparted to H by E" should be estimated in specific cases.

Since it is not at present our purpose to deal at length with the foundations of probability theory, while on the other hand the use of some notion of degree of confirmation in the vague sense introduced here seems to us unavoidable, we shall largely have to ignore the technical problems pointed out above. For practical purposes this does not mean that we shall maintain the fiction of there being a well-defined formula available that permits computation of $dc(H,E)$ for all H and E. Rather we shall assume, in specific cases arising in situations of interest, that reasonable and knowledgeable persons, when confronted with the question of ascertaining a value of $dc(H,E)$, will find this definitely, if vaguely, meaningful and will arrive at estimates of the value that will not be too widely disparate. This leads us to the next probabilistic concept that must be discussed.

Personal Probability   Personal, or subjective, probability is a measure of a person's confidence in, or subjective conviction of, the truth of some hypothesis. According to Savage,[9] it is measured behavioristically in terms of the person's betting behavior. If a person thinks that H is just about as likely as its negation $\sim H$, then, if he were placed in a situation where he had to make an even bet on either H or $\sim H$, he would presumably be indifferent to this choice. Similarly, if he thought H to be twice as likely as

9. L. J. Savage, *The Foundations of Statistics* (New York, 1952).

$\sim H$, he would have no preference as to which side to take in a $1:2$ bet on $H:\sim H$. Generalizing this idea, we shall say that the person attaches the personal probability p to the hypothesis H if he is found to be indifferent between the choice of receiving, say, one dollar if H turns out to be true or receiving $p/(1-p)$ dollars if H turns out to be false (his "personal expectation" in either case being p dollars).[10]

We shall call a person "rational" if (1) his preferences (especially with regard to betting options) are mutually consistent or at least, when inconsistencies are brought to his attention, he is willing to correct them; (2) his personal probabilities are reasonably stable over time, provided he receives no new relevant evidence; (3) his personal probabilities are affected (in the right direction) by new relevant evidence; and (4) in simple cases where the evidence E at his disposal is known, and E and H are such that dc (H,E) is defined, his personal probability regarding H is in reasonable agreement with the latter; in particular, he is indifferent as to which side to take in a bet that to his knowledge is a "fair" bet.

A (predictive) "expert" in some subject matter is a person who is *rational* in the sense discussed, who has a large background knowledge E in that field, and whose predictions (actual or implicit in his personal probabilities) with regard to hypotheses H in that field show a good record of comparative success in the long run. This is very much a relative concept, as it depends on the predictive performance of which the average nonexpert in the field would be capable. (In a temperate climate, a lay predictor can establish an excellent record by always forecasting good weather, but this would not support a claim to meteorological expertise.) Below we will give more detailed attention to predictive expertise.

10. There are certain well-known difficulties connected with this behavioristic approach, which we will ignore here. We will merely mention that in experimental situations designed to elicit personal probabilities, care must be taken that the stakes involved are in a range where the utility of money is effectively linear and the utility (or disutility) of gambling is negligible.

With regard to the relationship between degree of confirmation and personal probability, it may be said that $dc(H, E)$ is intended to be a conceptual reconstruction of the personal probability that an entirely rational person would assign to H, given that his entire relevant information is E. In practice this relation can be applied in both directions: In simple cases where we have a generally acceptable definition of "dc" we may judge a person's rationality by the conformity of his personal probabilities — or of his betting behavior — with computable (or, if his information E is uncertain, estimable) dc values. Conversely, once a person has been established as rational and possibly even as an expert in a field, we may use his personal probabilities as estimates, on our part, of the degrees of confirmation that should be assigned to given hypotheses.

We shall make use of these probability concepts presently, primarily in connection with the use of expert judgment for the predictive purposes. But we must first consider the use of evidence in prediction, beginning with some examples to illustrate the problems arising in the predictive use of probabilistic evidence.

7. Some Examples of the Use of Evidence in Prediction

The simplest use of evidence occurs when there is a direct reference to prior instances. Will my car start on this cold morning? Its record of successful starts on previous cold mornings is about 50 per cent. I would be unduly hopeful or pessimistic in assigning as personal probability of its starting today a number significantly different from one half. This use of a record of past instances as a basis for probability assignments with regard to future events is a common, and generally justified, inductive procedure (and of course is the basis on which a definition of degree of confirmation is constructed). However, under some circumstances it is a very poor way indeed of marshaling evidence.

Consider the case of Smith, who has been riding the bus to work for a year for a fare of 10 cents. One morning he is required to pay 15 cents. Smith may wonder if his return fare that evening will be 10 cents; but it is highly unlikely, despite the great preponderance of 10-cent rides in Smith's sample. For Smith well knows that public transportation fares do change, and not by whim but by adoption of a new fare structure. In the light of this item of *background information*, it is unreasonable for Smith to base his personal probability directly on the cumulative record of past instances.

This illustrates the need for the use of background knowledge as indirect evidence, in the sense of furnishing other than direct-instance confirmation. This need is encountered constantly in the use of evidence, and it constitutes one of the prime obstacles to a more sophisticated definition of degree of confirmation than has hitherto been achieved. Consider another example. Will my new neighbor move away again within 5 years? He is a carpenter (the average carpenter moves once every 10 years) and a bachelor (the average bachelor moves once every 3 years). I can assess the likelihood of my neighbor's moving within the next 5 years relative to either the reference class of carpenters or that of bachelors. Which one I should choose, or what weight I should give to each, must depend strongly on my background information as to the relative relevance of occupation versus marital status as a determining factor in changes of domicile.

Such reference-class problems arise even with statistical information of the simplest kind. Consider a sample of 100 objects drawn at random from a population, with the following outcome as regards possession of the properties P and Q:

	has Q	has not-Q
has P	1	9
has not-P	89	1

Given this information, what is the probability that another object drawn from the population, which is known to have the property P, will also have the property Q? Should we use a value around .1 (since only 1 of 10 observed P's is a Q) or a value around .9 (since altogether 90 per cent of the observed sample has the property Q)? Here again, an expedient use of the statistical evidence before us must rely on background information, if any, regarding the relevance of P-ness to Q-ness. If we know that most Texans are rich and most barbers poor, and are given as the only specific item of information about a man by the name of Jones that he is a Texan barber, we would do well to assign a low probability to the statement that Jones is rich, precisely because occupation is known to us to be more relevant to financial status than is location.

8. The Role of Expertise in Prediction

The implication of the examples we have been discussing is that a knowledge about past instances or about statistical samples, while indeed providing valuable information, is not the sole and sometimes not even the main form of evidence in support of rational assignments of probability values. In fact the evidential use of such *prima facie* evidence must be tempered by reference to background information, which frequently may be intuitive in character and have the form of a vague recognition of underlying regularities, such as analogies, correlations, or other conformities, the formal rendering of which would require the use of predicates of a logical level higher than the first.

The consideration of such underlying regularities is of special importance for the inexact sciences, particularly (but not exclusively) the social sciences,[11] because in this sphere we are constantly faced with situations in which statistical information

11. Use of background information to temper the application of statistical information is just as operative in the physical sciences — in engineering, for example; therefore, no difference in principle is involved here.

matters less than knowledge of regularities in the behavior of people or in the character of institutions, such as traditions and customary practices, fashions and mores, national attitudes and climates of opinion, institutional rules and regulations, group aspirations, and so on. For instance, in assessing the chances of a Republican presidential victory in 1980, a knowledge of the record of past election successes matters less than an insight into current trends and tendencies; or in answering a question as to the likelihood, say, of U.S. recognition of Communist China by 1979, it is hard to point to any relevant statistical evidence, yet there exists a mass of relatively undigested but highly relevant background information.

This nonexplicitness of background knowledge, which nonetheless may be significant or even predominantly important, is typical of the inexact sciences, as is the uncertainty as to the evidential weight to be accorded various pieces of *prima facie* information in view of indirect evidence provided by underlying regularities. Hence the great importance that must be attached to experts and to expertise in these fields. For the expert has at his ready disposal a large store of (mostly inarticulated) background knowledge and a refined sensitivity to its relevance, through the the intuitive application of which he is often able to produce trustworthy personal probabilities regarding hypotheses in his area of expertness.

The important place of expert judgment for predictions in the inexact sciences is further indicated by the prominence of quasi-laws among the explanatory instrumentalities of this domain. Since the conditions of applicability of such generalizations are neither fully nor even explicitly formulable, their use in specific circumstances presupposes the exercise of sound judgment as to their applicability to the case at hand. The informed expert, with his resources of background knowledge and his cultivated sense of the relevance and bearing of generalities in particular cases, is best able to carry out the application of quasi-laws necessary for reasoned prediction in this field.

9. The Problem of the Predictive Use of Evidence in an Inexact Context

In summary, the foregoing illustrations of the predictive use of evidence may be said to indicate that we are frequently confronted with what must be considered as a problematical, and far from ideal, epistemological situation. The examples we have been considering show that in assessing the probability of a hypothesis H — typically a description of some future event — we are in many instances required to rely not merely on some specific and explicit evidence E, but also on a vast body of potentially relevant background knowledge K, which is in general not only vague in its extent (and therefore indefinite in content) but also deficient in explicit articulation. In many practical applications, particularly in the inexact sciences, not even that part of K that is suitably relevant to H can be assumed to be explicitly articulated, or even articulable. One is unable to set down in sentential form everything that would have to be included in a full characterization of one's knowledge about a familiar room; and the same applies equally, if not more so, to a political expert's attempt to state all he knows that might be relevant to a question such as that of U.S. recognition of Communist China.

These considerations point up a deficiency for present purposes in the usual degree-of-confirmation concept quite apart from those already mentioned. For such an indefinite K, we cannot expect $dc(H, E \& K)$ to be determinable or even defined. This suggests, as a first step, the desirability of introducing a concept $dc_K(H, E)$ the "degree of confirmation of H on E in view of K" — which is defined to be equal to $dc(H, E \& K)$ whenever it is possible to articulate K fully within the same language in which H and E are stated. But how is such a quantity to be determined when K is not fully formulated? Furthermore, in addition to the difficulty involved in formulating it completely, K almost invariably contains probability statements (both of an objective, or dc, type, and of the indirect form "So-and-so attaches to H the personal probability p"). To date, there is no hint of any suggestion as to how "$dc(H, X)$" might be formally defined when X contains incompletely formulated matter.

Faced with this situation, which is surely not likely to be resolved in the near future, we must either for the present renounce all claims to systematized prediction in the inexact sciences, or, as indicated earlier, turn to unorthodox methods that are based on judicious and systematic reliance on expert judgment. One such course, to which we previously alluded, may possibly help us out of the present perplexity. Let A be an expert and $K(A)$ his relevant background knowledge. Then A's personal probability, $pp_A(H, E)$, may be taken as an estimate on *our* part of $dc_{K(A)}(H, E)$. Thus the device of using the personal probabilities of experts, extracted by appropriately devised techniques of interrogation, can serve as a means for measuring quantities of the dc type even in cases where there is no hope of applying the formal degree-of-confirmation concepts.

It might seem that in resorting to this device we conjure up a host of new problems, because—to all appearances—we are throwing objectivity to the winds. Of course, since we insist on remaining within our own definition of scientific activity, we do not propose to forego objectivity. However, before attempting to analyze the possibility of salvaging objectivity in this situation, it may be well to look at a few examples illustrating the application of expertise in the sense just described.

10.   The Intrinsic Use of Experts for Prediction

A source of characteristic examples of the predictive use of expert judgment is provided by the field of diagnostics, especially medical diagnostics.[12] A patient, let us assume, exhibits a pattern of symptoms such that it is virtually certain that he has either ailment A or ailment B, with respective probabilities of .4 and

12. An extensive and useful discussion of medical prediction is contained in P. E. Meehl's book on *Clinical vs. Statistical Prediction* (Minneapolis, 1954).

.6. where these probabilities derive from the statistical record of past cases. Thus the entire body of explicit symptomatic evidence is (by hypothesis) such as to indicate a margin in favor of the prediction that the patient suffers from disease B rather than A, and thus may respond positively to a corresponding course of treatment. But it is quite possible that an examining physician, taking into consideration not only the explicit indicators that constitute the symptoms (temperature and blood pressure, for example) but also an entire host of otherwise inarticulated background knowledge with regard to this particular patient (such as the circumstances of the case) may arrive at a diagnosis of disease A rather than B. Thus the use of background information, in a way that is not systematized but depends entirely on the exercise of informal expert judgment, may appropriately lead to predictive conclusions in the face of *prima facie* evidence that points in the opposite direction.

Quite similar in its conceptual structure to the foregoing medical example are various other cases of predictive expertise in the economic sphere. The advice of an expert investment counselor, for example, may exhibit essentially the same subtle employment of nonarticulate background knowledge that characterized the prediction of the diagnostician.

Again, in such essentially sociological predictions of public reactions as are involved in the advertising and marketing of commercial products, the same predictive role of expert judgment comes into play. When the production of a motion picture is completed, a decision must be made regarding the number of prints to be made. There are economic reasons for an accurate prediction of the need: if too few prints are ready to meet the immediate demand, film rental income will be lost; on the other hand, the prints are costly, and an oversupply leads to considerable excess expenditure. Here again, as in the medical or economic examples, certain limited predictions can be based wholly on the record of past statistics in analogous instances. The presence of certain actors in the cast, the topic, the theme and setting of the film, perhaps even its reception by preview audiences, may suggest a probability distribution for its demand. However, the major studios involved in motion-picture production are not content to rely on these explicit indicators alone. They are aware of the potential influence of a host of subtle intangibles such as so-called audience appeal, of the dependence on competitive offerings, and of other factors that are difficult to predict such as timeliness with reference to current events—all of which are susceptible of explicit statistical treatment only with the greatest difficulty, if at all. Therefore they prudently rely on the forecasts of professional experts in the field, who have exhibited a demonstrated ability to supplement the various explicit elements by appropriate use of their capacities for an intuitive appraisal of the many intangible factors that critically affect the final outcome.

Other examples drawn from the applied sciences, engineering, industry, politics will easily suggest themselves. What they have in common is the reliance, in part or wholly, on an expert, who here functions in an intrinsic rather than an extrinsic role. By extrinsic expertise we mean the kind of inventiveness, based on factual knowledge and the perception of previously unnoticed relationships, that goes into the hypothesizing of new laws and the construction of new theories; it is, in other words, the successful activity of the scientist qua scientist. Intrinsic expertise, by contrast, is not invoked until after a hypothesis has been formulated and its probability, in the sense of degree of confirmation, is to be estimated. The expert, when performing intrinsically, thus functions within a theory rather than on the theory-constructing level.

11. The Role of Prediction as an Aid to Decision-Making

The decisions that professional decision-makers—governmental administrators, company presidents, military commanders, and so on—are called on to make inevitably turn on the question of future developments, since their directives as to present actions are invariably conceived with a view to future results. Thus a reliance on predictive ability is nowhere more overt and more pronounced than in the area of policy formation, and decision-making in general.

For this reason, decision-makers surround themselves with staffs of expert advisers, whose special knowledge and expertise must generally cover a wide field. Some advising experts may have a great store of factual knowledge, and can thus serve as walking reference books. Others may excel through their diagnostic or otherwise predictive abilities. Still others, such as operations analysts and management consultants, may have a special analytical capacity to recognize the structure of the problems at hand, thus aiding in the proper use of the contributions of the other two types of experts. The availability of such special expertise constitutes for the decision-maker a promise of increased predictive ability essential to the more effective discharge of his own responsibilities, and the ultimate function of expert advice is almost always to make a predictive contribution.

While the dependence of the decision-makers on expert advisers is particularly pronounced in social-science contexts—in the formulation of economic and political policies, for example— such dependence on expertise ought by no means to be taken to contradistinguish the social from the physical sciences. In certain engineering applications, particularly of relatively underdeveloped branches of physics (such as the applied physics of extremes of temperature or velocity) the reliance on know-how and expert judgment is just as pronounced as it is in the applications of political science to foriegn-policy formation. The use of experts for prediction does *not* constitute a line of demarcation between the social and the physical sciences, but rather between the exact and the inexact sciences. And, as we have already said, certain areas of applied physics are in the present state of our knowledge just as inexact as much of the social sciences, if indeed not more so.

Although we have held that the primary service of expert advisers to decision-makers is in functioning as "predictors," we by no means intend to suggest that they act as fortunetellers, trying to foresee specific occurrences for which the limited intellectual vision of the nonexpert is insufficient. For the decision-supporting uses of predictive expertise, there is in general no necessity for an anticipation of particular future occurrences. It suffices that the expert be able to sketch out adequately the general directions of future developments, to anticipate — as we have already suggested — some of the major critical junctures ("branch points") on which the course of these developments will hinge, and to make contingency predictions with regard to the alternatives associated with them.

Whereas the value of scientific prediction for sound decision-making is beyond question, it can hardly be claimed that the inexact sciences have the situation regarding the use of predictive expertise well in hand. Quite to the contrary, available evidence suggests that significant improvements are possible in the predictive instruments available to the decision-maker. These improvements are contingent on the development of methods for the more effective predictive use of expert judgment. In our final sections we shall give consideration to some of the problems involved in this highly important but hitherto largely unexplored area.

12. Justification of the Intrinsic Use of Expertise

Let us return to the problem of preserving objectivity in the face of reliance on expertise. Can we accept the use of intrinsic expert judgment within the framework of an inductive procedure without laying ourselves open to the charge of abandoning objective scientific methods and substituting rank subjectivity?

To see that explicit use of expert judgment is not incompatible with scientific objectivity, let us look once more at the medical-diagnosis example of the preceding section. Consider the situation in which a diagnostician has advised that a patient be treated for ailment *A* (involving, say, a major surgical operation) rather than *B* (which might merely call for a special diet). Our willingness, in this case, to put our trust in the expert's judgment surely would not be condemned as an overly subjective attitude. The reasons why our reliance on the expert is objectively justified are not difficult to see. For one thing, the selection of appropriate experts is not a matter of mere personal preference but is a procedure governed by objective criteria (about which more will be said in the ensuing section). But most importantly, the past diagnostic performance record makes the diagnostician an objectively reliable indicator (of diseases), in the same sense that one of any two highly correlated physical characteristics is an indicator of the other. ("If most hot pieces of iron are red, and vice versa, and if this piece of iron is red, then this piece is probably hot.")

Even if the expert's explicit record of past performance is unknown, reliance on his predictions may be objectively justified on the basis of general background knowledge concerning his credentials as an expert. The objective reliability of experts' pronouncements may also be strongly suggested by the fact that they often exhibit a high degree of agreement with one another, which—at least if we have reason to assume the pronouncements to be independent—precludes subjective whim.

Epistemologically speaking, the use of an expert as an objective indicator, as illustrated by the example of the diagnostician, amounts to considering the expert's predictive pronouncement as an integral, intrinsic part of the subject matter, and treating his reliability as a part of the theory about the subject matter. Our information about the expert is conjoined to our other knowledge about the field, and we proceed with the application of precisely the same inductive methods that we would apply in cases where no use of expertise is made. Our "data" are supplemented by the expert's personal probability valuations and by his judgments of relevance (which, by the way, could be derived from suitable personal probability statements), and our "theory" is supplemented by information regarding the performance of experts.

In this manner the incorporation of expert judgment into the structure of our investigation is made subject to the *same* safeguards that are used to assure objectivity in other scientific investigations. The use of expertise is therefore no retreat from objectivity or reversion to an idiosyncratic reliance on subjective taste.

13.   Criteria for the Selection of Predictive Experts

The first and most obvious criterion of expertise is of course knowledge. We resort to an "expert" precisely because we expect his information and the body of experience at his disposal to constitute an assurance that he will be able to select the needed items of background information, determine the character and extent of their relevance, and apply these insights to the formulation of the required personal probability judgments.

However, the expert's knowledge is not enough; he must be able to bring it to bear effectively on the predictive problem at hand, and this not every expert is able to do. It becomes necessary also to place some check on his predictive efficacy and to take a critical look at his past record of predictive performance.

The simplest way to score an expert's performance is in terms of "reliability": his *degree of reliability* is the relative frequency of cases in which, when confronted with several alternative hypotheses, he ascribed to the eventually correct alternative among them a greater personal probability than to the others.

This measure, though useful, must yet be taken with a grain of salt, for there are circumstances in which even a layman's degree of reliability, as defined above, can be very close to 1. For instance, in a region of very constant weather, a layman can prognosticate the weather quite successfully by always predicting the same weather for the next day as for the current one. Similarly, a quack who hands out bread pills and reassures his patients of recovery "in due time" may prove right more often than not and yet have no legitimate claim to being classified as a medical expert. Thus what matters is not so much an expert's absolute degree of reliability but his relative degree of reliability, that is, his reliability as compared with that of the average person. But even this may not be enough. In the case of the medical diagnostician discussed earlier, the layman may have no information that might give him a clue as to which of diseases A and B is the more probable, whereas anyone with a certain amount of rudimentary medical knowledge may know that disease A generally occurs much more frequently than disease B; yet his prediction of A rather than B on this basis alone would not qualify him as a reliable diagnostician. Thus a more subtle assessment of the qualifications of an expert may require his comparison with the average person having some degree of general background knowledge in his field of specialization. One method of scoring experts somewhat more subtly than just by their reliability is in terms of their "accuracy": the *degree of accuracy* of an expert's predictions is the correlation between his personal probabilities p and his correctness in the class of those hypotheses to which he ascribed the probability p. Thus of a highly accurate predictor we expect that of those hypotheses to which he ascribes, say, a probability of 70 per cent, approximately 70 per cent will eventually turn out to be confirmed. Accuracy in this sense, by the way, does not guarantee reliability,[13] but accuracy in addition

13. For instance, suppose experts A and B each gave 100 responses, assigning probabilities .2, .4, .6, and .8 to what in fact were the correct alternatives among 100 choices of H and $\sim H$, as follows:

p	A	B
.2	10	0
.4	20	0
.6	30	60
.8	40	40

to reliability may be sufficient to distinguish the real expert from the specious pretender.

14.   The Dependence of Predictive Performance on Subject Matter

Not only are some experts better predictors than others, but subject-matter fields differ from one another in the extent to which they admit of expertise. This circumstance is of course in some instances owing to the fact that the scientific theory of the field in question is relatively undeveloped. The geology of the moon and the meteorology of Mars are less amenable to prediction than their mundane counterparts, although no greater characteristic complexity is inherent in these fields. In other cases, however, predictive expertise is limited despite a high degree of cultivation of a field, because the significant phenomena hinge on factors that are not particularly amenable to prediction.

In domains in which the flux of events is subject to gradual transitions and constant regularities (such as, say, astronomy), a high degree of predictive expertise is possible. In those fields, however, in which the processes of transition admit of sharp jolts and discontinuities, which can in turn be the effects of causal processes so complex and intricate as to be "chance" occurrences for all practical purposes, predictive expertise is inherently less feasible. The assassination of a political leader can altogether change the policies of a nation, particularly when such a nation does not have a highly developed complex of institutions to ensure gradualness in public affairs. Clearly no expert on a particular country can be expected to have the data needed to predict specific assassinations; that is, his relevant information

---

Then A is perfectly *accurate* (for example, exactly 60 per cent, or 30, of the 20 + 30, or 50, cases to which he assigned .6 were correct), but he is only 70 per cent *reliable*; B, on the other hand, is 100 per cent reliable, but his accuracy is quite faulty (for example, 100 per cent, rather than 60 per cent, of the 60 cases to which he assigned .6 were correct).

is virtually certain not to include the precisely detailed know-ledge of the state of mind of various key figures that might give him any basis whatsoever for assigning a numerical value as his personal probability to the event in question. This situation is quite analogous to that of predicting the outcome of a particular toss of a coin; only the precise dynamic details of the toss's initial conditions might provide a basis for computing a probab-ility other than one half for the outcome, and these details again are almost certainly unavailable. We may here legitimately speak of "chance occurrences," in the sense that an expert, unless he has the most unusual information at his disposal, is in no better a position than the layman to make a reliable prediction.

In the inexact sciences, particularly in the social sciences, the critical causal importance of such chance events makes predictive expertise in an absolute sense difficult and sometimes impossible, and it is this, rather than the quality of his theoretical machinery, that places the social scientist in a poor competitive position relative, say, to the astronomer.

However, when the expert is unable to make precise predictions because of the influence of chance factors, he can at very least indicate the major contingencies on which future developments will hinge. Even though the expert cannot predict the specific course of future events in an unstable country, he should be able to specify the major branch points of future contingencies, and to provide personal probabilities conditionally with respect to these. Thus, for example, although it would be unreasonable to expect an expert on the American economy to predict with pre-cision the duration of a particular phase of an economic cycle (such as a recession), it is entirely plausible to ask him to specify the major potential "turning points" in the cycle (such as increas-ed steel production at a certain juncture), and to indicate the probable courses of development ensuing upon each of the specified alternatives.

Such differences in predictability among diverse subject-matter fields lead to important consequences for the proper use of experts. One obvious implication is that it may clearly be more profitable to concentrate limited resources of predictive expertise on those portions of a broader domain that are inherently more amenable to prediction. For example in a study of long-range political developments in a particular geographic area, it might in some cases be preferable to focus on demographic developments rather than the evolution of programs and platforms of political parties.

However, the most important consideration is that even in subject-matter fields in which the possibility of prediction is very limited, the exercise of expertise, instead of being applied to the determination of *absolute* personal probabilities with respect to certain hypotheses, ought rather more profitably be concentrated on the identification of the relevant branch points and the associated problem of the *relative* personal probabilities for the hypothesis in question, relative, that is, to the alternatives arising at these branch points.

Even in predictively very "difficult" fields — such as the question of the future foreign policy of an unstable country — the major branch points of future contingencies are frequently few enough for actual enumeration, and although outright prediction cannot be expected, relative predictions hinging on these principal alternative contingencies can in many instances serve the same purposes for which absolute predictions are ordinarily employed. For example, it would be possible for a neighboring state, in formulating its own policy toward this country, to plan not for "the" (one and only) probable course of developments but to design several policies, one for each of the major contingencies, or perhaps even a single policy that could deal effectively with all the alternatives.

15. Predictive Consensus Techniques

The predictive use of an expert takes place within a rationale which, on the basis of our earlier discussion, can be characterized as follows: We wish to investigate the predictive hypothesis H; we fix, with the expert's assistance, on the major items of the body of explicit evidence E that is relevant to this hypothesis; we then use the expert's personal probability valuation $pp(H,E)$ as *our* estimate of the degree of confirmation of H on the basis of E, that is, as our estimated value of $dc(H,E)$.

This straightforward procedure, however, is no longer adequate in those cases in which *several* experts are available. For here we have not the single value $pp(H,E)$ of only one expert, but an entire series of values, one for each of the experts: $pp_1(H,E)$, $pp_2(H,E)$, and so on. The problem arises: How is the best joint use of these various expert valuations to be made?[14]

Many possible procedures for effecting a combination among such diverse probability estimates are available. One possibility, and no doubt the simplest, is to select one "favoured" expert, and to accept his sole judgment. We might, for example, compare the past predictive performance of the various experts, and select that one whose record has been the most successful.

Another simple procedure is to pool the various expert valuations into an average of some sort, possibly the median, or a mean weighted so as to reflect past predictive success.

Again, the several experts might be made to act as a single group, pooling their knowledge in round-table discussion, if possible eliminating discrepancies in debate, and the group might then, on the basis of its corporate knowledge, be asked to arrive at one generally agreeable corporate "personal" probability as its consensus, which would now serve as our dc estimate. (A weakness in this otherwise very plausible-sounding procedure is that the consensus valuation might unduly reflect the views of the most respected member of the group, or of the most persuasive.)

One variant of this consensus procedure is to require that the experts, after pooling their knowledge in discussion, and perhaps after debating the issues, set down their separate "second-guess"

14. Compare A. Kaplan, A. L. Skogstad, and M. A. Girshick, "The Prediction of Social and Technological Events," *Public Opinion Quarterly*, Spring, 1950.

personal probabilities, revising their initial independent valuations in the light of the group work. These separate values are then combined by an averaging process to provide our dc estimate. The advantage of such a combination of independent values over the use of a single generally acceptable group value is that it tends to diminish the influence of the most vociferous or influential group member. Incidentally, in any consensus method in which separate expert valuations are combined, we can introduce the refinement of weighting an expert's judgment so as to reflect his past performance.

Another consensus procedure, sometimes called the "Delphi Technique," eliminates committee activity altogether, thus further reducing the influence of certain psychological factors, such as specious persuasion, the unwillingness to abandon publicly expressed opinions, and the bandwagon effect of majority opinion. This technique replaces direct debate by a carefully designed program of sequential individual interrogations (best conducted by questionnaires) interspersed with information and opinion feedback derived by computed consensus from the earlier parts of the program. Some of the questions directed to the respondents may, for instance, inquire into the "reasons" for previously expressed opinions, and a collection of such reasons may then be presented to each respondent in the group, together with an invitation to reconsider and possibly revise his earlier estimates. Both the inquiry into the reasons and subsequent feedback of the reasons adduced by others may serve to stimulate the experts into taking into due account considerations they might through inadvertence have neglected, and to give due weight to factors they were inclined to dismiss as unimportant on first thought.

We have done no more here than to indicate some examples from the spectrum of alternative consensus methods. Clearly there can be no one universally "best" method. The efficacy of such methods obviously depends on the nature of the particular subject matter and may even hinge on the idiosyncrasies and personalities of the specific experts (such as their ability to work as a group). Indeed this question of the relative effectiveness of the various predictive consensus techniques is almost entirely an open problem for empirical research, and it is strongly to be hoped that more experimental investigation will be undertaken in this important field.

16.   Simulation and Pseudo-Experimentation

We have thus far, in discussing the intrinsic use of experts or of groups of experts, described their function as being simply to predict, in the light of their personal probabilities, the correctness or incorrectness of proposed hypotheses. This description, while apt in principle and often in practice, may prove for some instances to be greatly oversimplified.

For one thing, in situations concerned with complicated practical problems, no clear-cut hypothesis to which probability values could be meaningfully attached may be immediately discernible. For another, several kinds of expertise, all interacting with each other, may have to be brought to bear simultaneously in anything but a straightforward manner. Examples of such cases are provided by questions such as "How can tension in the Middle East be relieved?" "What legislation is needed to reduce juvenile delinquency?" "How can America's schools improve science instruction?" Here, before even a single predictive expert can be used intrinsically, some at least rudimentary theoretical framework must be constructed within which predictive hypotheses can be stated. This, of course, calls for expertise of the extrinsic kind. Generally, the process involved is somewhat as follows. The situation at hand (say, current crime patterns) is analyzed—that is, it is stated in terms of certain specific and, it is hoped, well-defined concepts. This step usually involves a certain amount of abstraction, in that some aspects of the situation that are judged irrelevant are deliberately omitted from the description. Then either some specific action is proposed, and a hypothesis stated as to its effects on the situation in question or, more typically, a law or quasi-law is formulated, stating that in situations of the kind at hand, actions of a certain kind will have such-and-such consequences.

A variant of this is of considerable epistemological significance. Instead of describing the situation directly, a *model* of it is constructed, which may be either mathematical or physical, in which each element making up the real situation is simulated by a mathematical or physical object, and its relevant properties and relations to other elements are mirrored by corresponding simulative properties and relations. For example, any geographical map may be considered a (physical) model of some sector of the world; the planetary system can be simulated mathematically by a set of mass-points moving according to Kepler's laws; and a city's traffic system can be simulated by setting up a miniature model of its road net, traffic signals, and vehicles.[15] Now, instead of formulating hypotheses and predictions directly about the real world, it is possible to do the same thing about the model. Any results obtained from an analysis of the model, to the extent that it truly simulates the real world, can then later be translated back into the corresponding statements about the latter. This injection of a model has the advantage that it admits of what may be called "pseudo-experimentation" ("pseudo" because the experiments are carried out in the model, not in reality). For example, in the case of traffic-system analysis, pseudo-experimentation may produce reliable predictions as to what changes in the time sequence of traffic signals will improve the flow of traffic through the city.

Pseudo-experimentation is nothing but the systematic use of the classical idea of a hypothetical experiment; it is applied when

15. On simulation in traffic research, see W. H. Glanville, "Road Safety and Traffic Research in Great Britain," *Journal of the Operations Research Society of America*, vol. 3, (1955), pp. 283–299, reprinted in J. F. McCloskey and J. M. Coppinger, *Operations Research for Management*, vol. 2 (Baltimore, 1956), pp. 82–100.

true experimentation is too costly or physically or morally impossible[16] or, as we shall discuss next, when the real-world situation is too complex to permit the intrinsic use of experts. The application of simulation techniques is a promising approach, the fruitfulness of which has only begun to be demonstrated in documented experiments.[17] It is particularly promising when it is desirable to employ intrinsically several experts with varying specialties in a context in which their forecasts cannot be entered independently but where they are likely to interact with one another. Here a model furnishes the experts with an artificial, simulated environment. Within this environment they can jointly and simultaneously experiment, responding to the changes induced by their actions and acquiring through feedback the insights necessary to make successful predictions within the model and thus indirectly about the real world.

This technique lends itself particularly to predictions regarding the behavior of human organizations, inasmuch as the latter can be simulated most effectively by having the experts play the roles

16. Classical examples of such pseudo-experimentation are found in atomic physics and in operations analysis. Atomic physics deals with particles so small as to preclude experimentation requiring direct manipulation and observation; what has been done here is to construct a mathematical model of certain atomic and nuclear processes and then to use Monte Carlo sampling techniques to conduct not real, but paper, experiments, in which the paths of fictitious particles of the model are "observed" as the latter go through a series of random collisions, deflections, or what not; in this "experiment" the features supposedly not under the control of the "experimenter" are assumed to be subject to given probability distributions, so that the chance fluctuations that would naturally occur can be simulated in the model by the operation of an artificial chance device. Similarly, the military, in order to evaluate the effectiveness of alternative weapon systems (which clearly cannot be fully tested directly), conduct pseudo-experiments on simulation models.

17. See, for instance, J. L. Kennedy and R. L. Chapman, *The Background and Implications of the Systems Research Laboratory Studies*, Symposium on Air Force Human Engineering, Publication No. 455, National Academy of Sciences, National Research Council (1956); and R. L. Chapman, "Simulation in RAND's Systems Research Laboratory," and W. W. Haythorn, "Simulation in RAND's Logistics Systems Laboratory," both in *Report of Symposium on System Simulation* (Baltimore, 1958).

of certain members of such organizations and act out what in their judgment would be the actions, in the situation simulated, of their real-life counterparts.[18] Generally it may be said that in many cases judicious pseudo-experimentation may effectively annul the oft-regretted infeasibility of carrying out experiments proper in the social sciences by providing an acceptable substitute that, moreover, has been tried and proved in the applied physical sciences.

17. Operational Gaming

A particular case of simulation involving role-playing by the intrinsic experts is known as operational *gaming*, a special case of which is war gaming. A simulation model may properly be said to be gaming a real-life situation if the latter concerns decisionmakers in a context involving conflicting interests. In operational gaming, the simulated environment is particularly effective in reminding the expert, in his role as a player, to take *all* the potentially relevant factors into account in making his predictions; for if he does not, and chooses a tactic or strategy that overlooks an essential factor, an astute "opponent" will soon enough teach him not to make such an omission again.

Aside from the obvious application of gaming to the analysis of military conflict—of which there are numerous examples, ranging from crude map exercises to sophisticated enterprises requiring the aid of high-speed computing equipment—gaming has been used to gain insights into the nature of political and economic conflict. In the political field, cold-war situations have been explored in this manner; and in the economic field, inroads have been made into analyses of bargaining and industrial competition.[19]

18. See, for example, the American Management Association study of *Top Management Decision Simulation* (New York, 1957).

19. Compare C. J. Thomas and W. L. Deemer, Jr., "The Role of Operational Gaming in Operations Research," *Journal of the Operations Research Society*

We note in passing that operational games differ greatly in the completeness of their rules. These may be complete enough so that at each stage the strategic options at the players' disposal are wholly specified and the consequences resulting from the joint exercise of these options are entirely determined; this would mean that the model represents a complete theory of the phenomena simulated in the game. On the other hand, neither of these factors may be completely determined by the rules, in which case it is up to an umpiring staff to allow or disallow proposed strategies and assess their consequences. Umpiring in this sense is yet another important device for the use of expertise intrinsically within the framework of a scientific theory (namely, the model in question).

18.   Review of the Main Theses

Before bringing this discussion to its conclusion, it is appropriate to pause briefly to review the main points of the foregoing deliberations. Our starting point was the distinction between the "exact" and the "inexact" areas of science. It is our contention that this distinction is far more important and fundamental from the standpoint of a correct view of scientific method than is the case with superficially more pronounced distinctions based on

---

of America (*JORSA*), vol. 5 (1957), pp. 1–27; A. M. Mood and R. D. Specht, *Gaming as a Technique of Analysis*, The RAND Corporation, Paper P-579, October 19, 1954; W. E. Cushen, "Operational Gaming in Industry," in J. F. McCloskey and J. M. Coppinger, *Operations Research for Management*, vol. 2 (Baltimore, 1956), pp. 358–375; R. Bellman *et al.*, "On the Construction of a Multi-stage, Multi-person Business Game," *JORSA*, vol. 5 (1957), pp. 469–503; F. M. Ricciardi, "Business War Games for Executives: A New Concept in Management Training," *Management Review* (1957), pp. 45–56; J. C. Harsanyi, "Approaches to the Bargaining Problem Before and After the Theory of Games," *Econometrica*, vol. 26 (1955); F. M. Ricciardi *et al.*, "Top Management Decision Simulation," *American Management Association* (1957); and G. K. Kalisch *et al.*, "Some Experimental *n*-Person Games," in Thrall, Coombs, and Davis (eds.), *Decision Processes* (New York, 1954).

subject-matter diversities, especially that between the social and the physical sciences. Some branches of the social sciences (certain parts of demography, for example), which are usually characterized by the presence of a formalized mathematical theory, are methodologically analogous to the exact parts of physics. By contrast, the applied, inexact branches of physical science—for instance, certain areas of engineering under "extreme" conditions—are in many basic respects markedly similar to the social sciences.

This similarity applies both to methods of explanation and to methods of prediction. Partly because of the absence of mathematically formalized theories, explanations throughout the area of the inexact sciences—within the physical- and the social-science settings alike—are apt to be given by means of the restricted generalizations that we have called "quasi-laws." The presence of such less-than-universal principles in the inexact sciences creates a different situation for the prospects of explanation and those of prediction in these fields. This suggests the desirability of developing the specifically predictive instrumentalities of these fields, for once the common belief in the identity of predictive and explanatory scientific procedures is seen to be incorrect, it is clearly appropriate to consider the nature and potentialities of predictive procedures distinct from those used for explanation. As for predictions in the inexact sciences (physical as well as social), these can be pragmatically acceptable (as a basis for action) when based on methodologically even less sophisticated grounds than are explanations—grounds such as expert judgment.

These general considerations regarding the methodology of the inexact sciences hold particularly intriguing implications for the possibility of methodological innovation in the social sciences. Here the possible existence of methods that are unorthodox in the present state of social-science practices merits the closest examination. This is particularly true with respect to the pragmatic applications of the social sciences (in support of decision-making, for example), in which the predictive element is preponderant over the explanatory.[20]

One consideration of this sort revolves about the general question of using expertise. We have stressed the importance in the social sciences of limited generalizations (quasi-laws), which cannot necessarily be used in a simple and mechanical way, but whose very application requires the exercise of expert judgment. And more generally, when interested in prediction in this area (especially for decision-making purposes), we are dependent on the experts' personal probability valuations for our guidance. A systematic investigation of the effective use of experts represents a means by which new and powerful instruments for the investigation of social-science problems might be forged.

Further, the use in social-science contexts of a variety of techniques borrowed from other, applied, sciences that are also inexact (such as engineering applications and military and industrial operations research) deserves the most serious consideration. For there are numerous possibilities for deriving leads on methods that are also potentially useful in the social sciences; in particular the use of simulation as a basis for conducting pseudo-experiments comes to mind. Finally, there is the important possibility of combining simulation with the intrinsic use of expertise, especially by means of the technique of operational gaming. This prospect constitutes a method whose potential for social-science research has hitherto been pretty much unexplored. It is our hope that this neglect will soon be remedied.

20. Two recent publications of particular value in this sphere are B. de Jouvenel, *The Art of Conjecture* (New York, 1967) and Olaf Helmer, *Social Technology* (New York, 1966). A comprehensive bibliography on predictive processes and futuristic methodology is given in K. Baier and N. Rescher (eds.), *Values and the Future* (New York, 1969).

BIBLIOGRAPHY ON SCIENTIFIC EXPLANATION

The theory of inductive reasoning and confirmation theory are excluded from the purview of this Bibliography. For the older literature of this area see:

Rudolf Carnap, *Logical Foundations of Probability* (Chicago, 1950; 2nd ed., 1960).

For the more recent literature see:

Henry E. Kyburg, "Recent Work in Inductive Logic," *American Philosophical Quarterly*, vol. 1 (1964), pp. 249–287.

I. General

A. *General Works on the Theory of Scientific Explanation*

1. BOOKS AND ANTHOLOGIES

Barker, S. F., *Induction and Hypothesis* (Ithaca, N.Y., 1957).
Bergmann, Gustav, *Philosophy of Science* (Madison, Wis., 1957).
Blanshard, Brand, *The Nature of Thought* (London, 1939) 2 vols.
Bradley, F. H., *The Principles of Logic* (London, 1922).
Braithwaite, R. B., *Scientific Explanation* (Cambridge and New York, 1953).
Broad, C. D., *Scientific Thought* (London, 1923).
Bunge, Mario, *Metascientific Queries* (Springfield, Ill., 1959).
Campbell, Norman R., *Physics, the Elements* (Cambridge, 1920).
———, *What Is Science?* (London, 1920; reprinted New York, 1952).
Caws, Peter, *The Philosophy of Science* (Princeton, N.J., 1965).
Craik, Kenneth J., *The Nature of Explanation* (Cambridge, 1943).
Danto, Arthur, and Sidney Morgenbesser (eds.), *Philosophy of Science* (Cambridge and New York, 1960).
Duhem, Pierre, *The Aim and Structure of Physical Theory*, tr. by P. P. Wiener (Princeton, N.J., 1953).
Feigl, Herbert, and May Brodbeck (eds.), *Readings in the Philosophy of Science* (New York, 1953).
Frank, Philipp, *Philosophy of Science* (Englewood Cliffs, N.J., 1962).
Hanson, N. R., *Patterns of Discovery* (Cambridge, 1958).
Harre, Romano, *An Introduction to the Logic of the Sciences* (London, 1960).
Hempel, Carl G., *Aspects of Scientific Explanation* (New York, 1965).
Hobson, E. W., *The Domain of Natural Science* (London, 1923).
Jevons, W. S., *The Principles of Science* (New York, 1900).
Kahl, Russell (ed.), *Studies in Explanation* (Englewood Cliffs, N.J., 1963).

Kaplan, Abraham, *The Conduct of Inquiry* (San Francisco, 1964).

Kemeny, John G., *A Philosopher Looks at Science* (New York, 1959).

Kuhn, T. S., *The Structure of Scientific Revolutions* (Chicago, 1963).

Lewis, C. I., *An Analysis of Knowledge and Valuation* (LaSalle, Ill., 1946).

Lindsay, Robert, and Henry Margenau, *Foundations of Physics* (New York, 1936).

Mach, Ernst, *The Science of Mechanics: A Critical and Historical Account of its Development*, tr. by T. J. McCormack, 2nd ed., ed. by C. S. Peirce (Chicago, 1902).

Madden, Edward H. (ed.), *The Structure of Scientific Thought: An Introduction to Philosophy of Science* (Boston, 1960).

Margenau, Henry, *The Nature of Physical Reality* (New York, 1950).

Mehlberg, Henryk, *The Reach of Science* (Toronto, 1958).

Meyerson, Emile, *Identity and Reality*, tr. by K. Loewenberg (London and New York, 1930).

Mill, John Stuart, *A System of Logic* (London and New York, 1947) originally published in 1843, Book VI, "The Logic of the Moral Sciences," abridged version in *John Stuart Mill's Philosophy of Scientific Method*, ed. by E. Nagel (New York, 1950).

Mises, Richard von, *Positivism: A Study in Human Understanding* (Cambridge, Mass., 1951).

Nagel, Ernest, *The Structure of Science: Problems in the Logic of Scientific Explanation* (New York, 1961).

Natanson, Maurice (ed.), *Philosophy of the Social Sciences: A Reader* (New York, 1963).

Northrop, F. S. C., *The Logic of the Sciences and the Humanities* (New York, 1947).

Pap, Arthur, *An Introduction to the Philosophy of Science* (New York, 1962).

Poincaré, Henri, *Science and Hypothesis* (New York, 1952).

Popper, Karl R., *Logik der Forschung* (Vienna, 1935). Translated as *The Logic of Scientific Discovery* (New York and London, 1959).

Reichenbach, Hans, *Experience and Prediction* (Chicago, 1938).

Scheffler, Israel, *The Anatomy of Inquiry* (London, 1953).

Schlesinger, George, *Method in the Physical Sciences* (New York, 1963).

Schlick, Moritz, *Philosophy of Nature*, tr. by A. von Zeppelin (New York, 1949). [On explanation see especially chap. 3.]

Stegmüller, Wolfgang, *Wissenschaftliche Erklärung und Begründung* (Berlin, Heidelberg, and New York, 1969).

Toulmin, Stephen, *The Philosophy of Science* (London, 1953).
————, *Foresight and Understanding* (Bloomington, Ind., 1961).
Whewell, William, *The Philosophy of the Inductive Sciences* (London, 1847).
Wiener, Philip (ed.), *Readings in the Philosophy of Science* (New York, 1953).

2. ARTICLES AND CHAPTERS

Barker, S. F., "The Role of Simplicity in Explanation" in H. Feigl and G. Maxwell (eds.), *Current Issues in the Philosophy of Science* (New York, 1961), and the "Comments" on this paper by W. Salmon, P. Feyerabend, and R. Rudner with Barker's "Rejoiners" thereto., pp. 265–285.
Brody, B. A., "Confirmation and Explanation," *The Journal of Philosophy*, vol. 65 (1968), pp. 282–299.
Bromberger, Sylvain, "Why Questions" in R. G. Colodny (ed.), *Mind and Cosmos* (Pittsburgh, 1966).
————, "An Approach to Explanation" in Ronald J. Butler (ed.), *Studies in Analytical Philosophy* (Oxford, 1965), pp. 72–105.
Ducasse, C. J., "Explanation, Mechanism, and Teleology," *The Journal of Philosophy*, vol. 23 (1926), pp. 150–155. Reprinted in H. Feigl and W. Sellars (eds.), *Readings in Philosophical Analysis* (New York, 1949).
Feigl, Herbert, "Some Remarks on the Meaning of Scientific Explanation," *The Psychological Review*, vol. 52 (1945), pp. 250–259. Reprinted in H. Feigl and W. Sellars (eds.), *Readings in Philosophical Analysis* (New York, 1949), pp. 510–514.
Feyerabend, P. K., "Explanation, Reduction, and Empiricism" in H. Feigl and G. Maxwell (eds.), *Minnesota Studies in the Philosophy of Science*, vol. 3 (Minneapolis, 1962), pp. 231–272.
Hayek, F. A., "Degrees of Explanation," *British Journal for the Philosophy of Science*, vol. 6 (1955–56), pp. 209–225.
Hempel, Carl G., "Deductive Nomological vs. Statistical Explanation" in H. Feigl and G. Maxwell (eds.), *Minnesota Studies in the Philosophy of Science*, vol. 3 (Minneapolis, 1962).
————, and Paul Oppenheim, "Studies in the Logic of Explanation," *Philosophy of Science*, vol. 15 (1948), pp. 135–175. Reprinted as "The Logic of Explanation" in H. Feigl and M. Brodbeck (eds.), *Readings in the Philosophy of Science* (New York, 1953), pp. 319–352.

Hesse, Mary B., "A New Look at Scientific Explanation," *Review of Metaphysics*, vol. 17 (1963), pp. 98–108.

Hospers, John, "On Explanation," *The Journal of Philosophy*, vol. 43 (1946), pp. 337–346.

——, "What is Explanation?" in A. Flew (ed.), *Essays in Conceptual Analysis* (London, 1956).

Kaplan, David, "Explanation Revisited," *Philosophy of Science*, vol. 28 (1961), pp. 429–436.

Kneale, William, "Induction, Explanation, and Transcendent Hypotheses" in W. Kneale, *Probability and Induction* (Oxford, 1949), pp. 92–110. Reprinted in H. Feigl and M. Brodbeck (eds.), *Readings in the Philosophy of Science* (New York, 1953), pp. 353–362.

Leach, James, "Dray on Rational Explanation," *Philosophy of Science*, vol. 33 (1966), pp. 61–69.

Mach, Ernst, "On the Economical Nature of Physical Inquiry" in T. J. McCormack (tr.), *Popular Scientific Lectures* (La Salle, Ill., 1943).

Miller, D. L., "Explanation vs. Description," *The Philosophical Review*, vol. 56 (1947), pp. 306–312.

Mischel, Theodore, "Pragmatic Aspects of Explanation," *Philosophy of Science*, vol. 33 (1966), pp. 40–60.

Morgenbesser, Sidney, "The Explanatory-Predictive Approach to Science" in B. Baumrin (ed.), *Philosophy of Science: The Delaware Seminar*, vol. 1 (New York, 1963).

Nagel, Ernest, "Introduction: Science and Common Sense" in E. Nagel, *The Structure of Science* (New York, 1961), pp. 1–14.

——, "Patterns of Scientific Explanation" in E. Nagel, *The Structure of Science* (New York, 1961), pp. 15–28.

Passmore, John, "Explanation in Everyday Life, in Science, and in History," *History and Theory*, vol. 2 (1962), pp. 105–123.

Poincaré, Henri, "Is Science Artificial?" in H. Poincaré, *The Value of Science* (New York, 1958, reprinted).

Scriven, Michael, "New Issues in the Logic of Explanation" in S. Hook (ed.), *Philosophy and History* (New York, 1963), pp. 339–361.

——, Review of E. Nagel, *The Structure of Science*, *Review of Metaphysics*, vol. 17 (1963–64), pp. 403–424.

Skarsgard, Lars, "Some Remarks on the Logic of Explanation," *Philosophy of Science*, vol. 25 (1958), pp. 199–207.

Smart, J. C. C., "Conflicting Views About Explanation" in R. S. Cohen and M. Wartofsky (eds.), *Boston Studies in the Philosophy of Science*, vol. 2 (New York, 1965).

————, "Philosophy and Scientific Plausibility" in P. Feyerabend and
  G. Maxwell (eds.), *Matter, Mind, and Method* (Minneapolis, 1966).
Weingartner, R. H., "Explanations and Their Justification," *Philosophy
  of Science*, vol. 28 (1961), pp. 300–305.
Yolton, John W., "Explanation," *British Journal for the Philosophy of
  Science*, vol. 10 (1959), pp. 194–208.

B. *The Deductive Model of Explanation and Its Critics*

1. BOOKS AND ANTHOLOGIES

Hanson, N. R., *Patterns of Discovery* (Cambridge, Mass., 1958).
Hempel, Carl G., *Aspects of Scientific Explanation* (New York, 1965).
Popper, Karl R., *The Logic of Scientific Discovery* (London, 1959).
Stegmüller, Wolfgang, *Wissenschaftliche Erklärung und Begründung*
  (Berlin, Heidelberg, and New York, 1969).

2. ARTICLES AND CHAPTERS

Ackermann, Robert, "Discussion: Deductive Scientific Explanation,"
  Philosophy of Science, vol. 32 (1965), pp. 155–167.
————, and Alfred Stenner, "Discussion: A Corrected Model of
  Explanation," *Philosophy of Science*, vol. 33 (1966), pp. 168–
  171.
Eberle, Rolf, David Kaplan, and Richard Montague, "Hempel and
  Oppenheim on Explanation," *Philosophy of Science*, vol. 28 (1961),
  pp. 418–428.
Feyerabend, P. K., "Explanation, Reduction, and Empiricism" in H.
  Feigl and G. Maxwell (eds.), *Minnesota Studies in the Philosophy
  of Science*, vol. 3 (Minneapolis, 1962), pp. 231–272.
Hempel, Carl G., "Deductive-Nomological vs. Statistical Explanation"
  in H. Feigl and G. Maxwell (eds.), *Minnesota Studies in the Philo-
  sophy of Science*, vol. 3 (Minneapolis, 1962), pp. 98–169.
————, "Explanation and Prediction by Covering Laws" in B. Baumrin
  (ed.), *Philosophy of Science: The Delaware Seminar*, vol. 1 (New
  York, 1963).
————, and Paul Oppenheim, "Studies in the Logic of Explanation,"
  Philosophy of Science, vol. 15 (1948), pp. 135–175. Reprinted as
  "The Logic of Explanation" in H. Feigl and M. Brodbeck (eds.),
  Readings in the Philosophy of Science (New York, 1953), pp.
  319–352.

Kim, Jaegwon, "On the Logical Conditions of Deductive Explanation," *Philosophy of Science*, vol. 30 (1963), pp. 286–291.

Nagel, Ernest, "The Deductive Pattern of Explanation" in E. Nagel, *The Structure of Science* (New York, 1961), pp. 29–46.

Newman, Fred, "Explanation Sketches," *Philosophy of Science*, vol. 32 (1965), pp. 168–172.

Scriven, Michael, "New Issues in the Logic of Explanation" in S. Hook (ed.), *Philosophy and History* (New York, 1963), pp. 339–361.

Stout, G. F., "Symposium: Mechanical and Teleological Causation," *Proceedings of the Aristotelian Society*, Supplementary Vol. 14 (1935), pp. 46–65.

C. *Probabilistic and Statistical Explanations*

1. BOOKS AND ANTHOLOGIES

Bohm, David, *Causality and Chance in Modern Physics* (London, 1957).

Born, Max, *Natural Philosophy of Cause and Chance* (London, 1949).

Carnap, Rudolf, *Logical Foundations of Probability* (Chicago, 1950).

Hempel, Carl G., *Aspects of Scientific Explanation* (New York, 1965).

Jeffreys, Harold, *Scientific Inference* (Cambridge, Mass., 1931).

Mises, Richard von, *Positivism: A Study in Human Understanding* (Cambridge, Mass., 1951).

Savage, L. J., *The Foundations of Statistics* (New York and London, 1954).

Stegmüller, Wolfgang, *Wissenschaftliche Erklärung und Begründung* (Berlin, Heidelberg, and New York, 1969).

2. ARTICLES AND CHAPTERS

Carlsson, Gösta, "Sampling, Probability and Causal Inference," *Theoria*, vol. 18 (1952), pp. 139–154.

Collins, Arthur W., "The Use of Statistics in Explanation," *British Journal for the Philosophy of Science*, vol. 17 (1966), pp. 127–140.

Gluck, S. E., "Do Statistical Laws Have Explanatory Efficacy?", *Philosophy of Science*, vol. 22 (1955), pp. 34–38.

Hempel, Carl G., "Deductive-Nomological vs. Statistical Explanation" in H. Feigl and G. Maxwell (eds.), *Minnesota Studies in the Philosophy of Science*, vol. 3 (Minneapolis, 1962), pp. 98–169.

————, "Maximal Specificity and Lawlikeness in Probabilistic Explanation," *Philosophy of Science*, vol. 35 (1968), pp. 116–133.

Humphreys, W. C., "Statistical Ambiguity and Maximal Specificity," *Philosophy of Science*, vol. 35 (1968), pp. 112–115.

Lenzen, Victor, "Philosophical Problems of the Statistical Interpretation of Quantum Mechanics," *Proceedings of the Second Berkeley Symposium on Mathematical Statistics and Probability* (Berkeley, 1951).

Massey, G. J., "Hempel's Criterion of Maximal Specificity," *Philosophical Studies*, vol. 19 (1968), pp. 43–47.

Northrop, F. S. C., "The Philosophical Significance of the Concept of Probability in Quantum Mechanics," *Philosophy of Science*, vol. 3 (1936), pp. 215–232. Reprinted in F. S. C. Northrop, *The Logic of the Sciences and the Humanities* (New York, 1947).

Novak, Stefan, "Some Problems of Causal Interpretation of Statistical Relationships," *Philosophy of Science*, vol. 27 (1960), pp. 23–38.

Peirce, Charles Sanders, "The Doctrine of Necessity Examined," *The Monist*, vol. 2 (1892), pp. 321–337. Reprinted in C. Hartshorne and P. Weiss (eds.), *Collected Papers of Charles Sanders Peirce*, vol. 6 (Cambridge, Mass., 1935).

Salmon, Wesley C., "The Status of Prior Probabilities in Statistical Explanation," *Philosophy of Science*, vol. 32 (1965), pp. 137–146.

Schlick, Moritz, "Die Kausalität in der gegenwärtigen Plysik," *Die Naturwissenschaften*, vol. 19 (1931), pp. 145–162.

Suppes, Patrick, "Probabilistic Inference and the Concept of Total Evidence" in J. Hintikka and P. Suppes (eds.), *Aspects of Inductive Logic* (Amsterdam, 1966).

D. *Studies in the History of Scientific Explanation*

1. BOOKS AND ANTHOLOGIES

Burtt, E. A., *The Metaphysical Foundations of Modern Physical Science* (New York, 1955).

Butterfield, Herbert, *The Origins of Modern Science, 1300–1800* (London, 1949).

Cohen, M. R., and I. R. Drabkin (eds.), *A Source Book in Greek Science* (New York, 1948).

Crombie, A. C., *Augustine to Galileo: The History of Science A.D. 400–1650* (London, 1952; Cambridge, Mass., 1953).

Duhem, Pierre, *The Aim and Structure of Physical Theory*, tr. by Philip P. Wiener (Princeton, 1954).

Hall, A. R., *The Scientific Revolution, 1500–1800* (New York, 1954).

Herschel, John F. W., *Preliminary Discourse on the Study of Natural Philosophy* (London, 1842).

Kahl, Russell (ed.), *Studies in Explanation: A Reader in the Philosophy of Science* (Englewood Cliffs, N.J., 1963).

Kuhn, T. S., *The Copernican Revolution* (New York, 1959).

Mach, Ernst, *The Science of Mechanics: A Critical and Historical Account of Its Development*, tr. by T. J. McCormack, 2nd. ed., by C. S. Peirce (Chicago, 1902).

Madden, Edward H. (ed.), *Theories of Scientific Method from the Renaissance to the Nineteenth Century* (Seattle, 1959).

Reichenbach, Hans, *The Rise of Scientific Philosophy* (Berkeley, 1951).

Whewell, William, *History of the Inductive Sciences* (London, 1857).

Whittaker, Edmund T., *From Euclid to Eddington* (Cambridge, 1949).

2. ARTICLES AND CHAPTERS

Butts, Robert E., "Hypothesis and Explanation in Kant's Philosophy of Science," *Archiv für Geschichte der Philosophie*, vol. 43 (1961), pp. 153–170.

————, "Kant on Hypotheses in the 'Doctrine of Method' and *Logik*," *Archiv für Geschichte der Philosophie*, vol. 44 (1962), pp. 185–203.

Turner, Joseph, "Maxwell on the Method of Physical Analogy," *British Journal for the Philosophy of Science*, vol. 6 (1955), pp. 226–238.

Zilsel, Edgar, "The Genesis of the Concept of Physical Law," *The Philosophical Review*, vol. 51 (1942), pp. 3–14.

II. Modes of Explanation

A. *Causal Explanation*

1. BOOKS AND ANTHOLOGIES

Ayer, A. J., *The Foundations of Empirical Knowledge* (London, 1940).

Bergmann, Gustav, *The Philosophy of Science* (Madison, Wisc., 1957).

Born, Max, *Natural Philosophy of Cause and Chance* (London, 1949).

Broad, C. D., *The Mind and Its Place in Nature* (London, 1929).

Bunge, Mario, *Causality: The Place of the Causal Principle in Modern Science* (Cambridge, Mass., 1959).

Cassirer, Ernst, *Determinism and Indeterminism in Modern Physics* (New Haven, 1956).

Frank, Philip, *Das Kausalgesetz und seine Grenzen* (Vienna, 1932).

———, *Interpretations and Misinterpretations of Modern Physics* (Paris, 1938).

———, *Philosophy of Science* (Englewood Cliffs, N.J., 1957).

Goetlind, Erik, *Bertrand Russell's Theories of Causation* (Uppsala, 1952).

Hart, H. L. A., and A. M. Honore, *Causation in the Law* (London, (1959).

Mill, J. S., *A System of Logic* (London, 1893).

Pap, Arthur, *An Introduction to the Philosophy of Science* (New York, 1962).

Reichenbach, Hans, *Philosophic Foundations of Quantum Mechanics* (Berkeley, 1944).

Schlick, Moritz, *Philosophy of Nature* (New York, 1949).

Silberstein, Ludwik, *Causality* (London, 1933).

Stegmüller, Wolfgang, *Wissenschaftliche Erklärung und Begründung* (Berlin, Heidelberg, and New York, 1969).

2. ARTICLES AND CHAPTERS

Albert, Ethel M., and Lewis Feuer, "Causality in the Social Sciences," *The Journal of Philosophy*, vol. 51 (1954), pp. 695–705.

Bohr, Neils, "Causality and Complementarity," *Philosophy of Science*, vol. 4 (1937), pp. 289–298.

Broad, C. D., "Mechanical Explanation and Its Alternatives," *Proceedings of the Aristotelian Society*, vol. 19 (1918), pp. 86–124.

Burks, A. W., "The Logic of Causal Propositions," *Mind*, vol. 60 (1951), pp. 363–382.

Fain, Haskell, "Some Problems of Causal Explanation," *Mind*, vol. 72 (1963), pp. 519–532.

Feigl, Herbert, "Notes on Causality" in H. Feigl and M. Brodbeck (eds.), *Readings in the Philosophy of Science* (New York, 1953), pp. 408–418.

Goudge, T. A., "Causal Explanation in Natural History," *The British Journal for the Philosophy of Science*, vol. 9 (1958), pp. 194–202.

Grünbaum, Adolf, "Causality and the Science of Human Behavior," *American Scientist*, vol. 40 (1952), pp. 665–676. Reprinted in H. Feigl and M. Brodbeck (eds.), *Readings in the Philosophy of Science* (New York, 1953), pp. 766–778.

Hartshorne, Charles, "Causal Necessities: An Alternative to Hume," *The Philosophical Review*, vol. 63 (1954), pp. 479–499.

Landé, Alfred, "Causality and Dualism on Trial" in B. Baumrin (ed.),

Philosophy of Science: The Delaware Seminar, vol. 1 (New York and London, 1963), pp. 327–351.

Mackie, J. L., "Counterfactuals and Causal Laws" in R. J. Butler (ed.), *Analytical Philosophy* (New York, 1962), pp. 66–80.

Margenau, Henry, "Physical Versus Historical Reality," *Philosophy of Science*, vol. 20 (1952), pp. 195–213.

Mayr, Ernst, "Cause and Effect in Biology," *Science*, vol. 134 (1961), pp. 1501–1506.

Nagel, Ernest, "The Causal Character of Modern Physical Theory" in E. Nagel, *Freedom and Reason* (New York, 1951). Reprinted in H. Feigl and M. Brodbeck (eds.), *Readings in the Philosophy of Science* (New York, 1953), pp. 419–437.

———, "Causality and Indeterminism in Physical Theory" in E. Nagel, *The Structure of Science* (New York, 1961), pp. 277–335.

Planck, Max, "The Concept of Causality in Physics" in *Scientific Autobiography and Other Papers*, tr. by F. Gaynor (New York, 1949). Reprinted in P. P. Wiener (ed.), *Readings in the Philosophy of Science* (New York, 1953).

Reichenbach, Hans, "Das Kausalproblem in der Physik," *Naturwissenschaften*, vol. 19 (1931), pp. 713–722.

Rescher, Nicholas, "The Stochastic Revolution and the Nature of Scientific Explanation," *Synthese*, vol. 14 (1962), pp. 200–215.

Riker, W. H., "Causes of Events," *The Journal of Philosophy*, vol. 55 (1958), pp. 281–291.

Russell, Bertrand, "The Notion of Cause" in *Mysticism and Logic* (New York, 1918). Reprinted in B. Russell, *Our Knowledge of the External World* (New York, 1929), pp. 245–256; reprinted as "On the Notion of Cause, with applications to the Free-Will Problem" in H. Feigl and M. Brodbeck (eds.), *Readings in the Philosophy of Science* (New York, 1953), pp. 387–407.

Schlick, Moritz, "Die Kausalität in der gegenwärtigen Physik," *Die Naturwissenschaften*, vol. 19 (1931), pp. 145–162.

———, "Causality in Everyday Life and in Recent Science" in H. Feigl and W. Sellars (eds.), *Readings in Philosophical Analysis* (New York, 1949).

Stegmüller, Wolfgang, "Das Problem der Kausalität," *Probleme der Wissenschaftstheorie. Festschrift für Victor Kraft* (Wien, 1960), pp. 171–190.

Ushenko, A. P., "The Principles of Causality," *The Journal of Philosophy*, vol. 50 (1953), pp. 85–101.

Waismann, Friedrich, "The Decline and Fall of Causality" in A. Crombie (ed.), *Turning Points in Physics* (Amsterdam, 1959), pp. 84–154.

Warnock, G. J., "Every Event Has a Cause" in A. Flew (ed.), *Logic and Language* (Oxford, 1955).

Weinberg, J. R., "The Idea of Causal Efficacy, *"The Journal of Philosophy*, vol. 47 (1950), pp. 397–407.

Workman, R. W., "Is Indeterminism Supported by Quantum Theory?", *Philosophy of Science*, vol. 26 (1959), pp. 251–259.

B. *Mechanical and Mechanistic Explanation*

1. BOOKS AND ANTHOLOGIES

D'Abro, A., *The Decline of Mechanism in Modern Physics* (New York, 1939).

Loeb, Jacques, *The Mechanistic Conception of Life* (Chicago, 1912).

2. ARTICLES AND CHAPTERS

Brandt, Richard, and Jaegwon Kim, "Wants as Explanations of Actions, *The Journal of Philosophy*, vol. 60 (1963), pp. 425–435.

Broad, C. D., "Mechanical Explanation and Its Alternatives," *Proceedings of the Aristotelian Society*, vol. 19 (1919), pp. 85–124.

———, C. A. Mace, G. F. Stout, and A. C. Ewing, "Symposium: Mechanical and Teleological Causation," *Proceedings of the Aristotelian Society*, supplementary vol. 14 (1935), pp. 83–112.

Dear, G. F., "Determinism in Classical Physics," *British Journal for Philosophy of Science*, vol. 11 (1961), pp. 289–304.

Ducasse, C. J., "Explanation, Mechanism, and Teleology," *The Journal of Philosophy*, vol. 23 (1926), pp. 150–155. Reprinted in H. Feigl and W. Sellars (eds.), *Readings in Philosophical Analysis* (New York, 1949).

Nagel, Ernest, "Mechanical Explanations and the Science of Mechanics" in E. Nagel, *The Structure of Science* (New York, 1961), pp. 153–202.

C. *Teleological Explanation*

1. BOOKS AND ANTHOLOGIES

Beckner, Morton, *The Biological Way of Thought* (Berkeley, 1968).

Cannon, Walter B., *The Wisdom of the Body* (New York, 1939).

Cohen, M. R., *Reason and Nature* (New York, 1931).
Emmet, Dorothy, *Function, Purpose and Powers* (New York and London, 1958).
MacIver, R. M., *Social Causation* (Boston, 1942).
Martindale, Don (ed.), *Functionalism in the Social Sciences*, Monograph No. 5 of the American Academy of Political and Social Science (Philadelphia, 1965).
Rudner, Richard S., *Philosophy of Social Science* (Englewood Cliffs, N.J., 1966).
Stegmüller, Wolfgang, *Wissenschaftliche Erklärung und Bergründung* (Berlin, Heidelberg, and New York, 1969).
Tolman, Edward C., *Purposive Behavior in Animals and Men* (New York, 1932).

2. ARTICLES AND CHAPTERS

Braithwaite, R. B., "Teleological Explanations: The Presidential Address," *Proceedings of the Aristotelian Society*, vol. 47 (1946–47), pp. i–xx.
Broad, C. D., "Mechanistic Explanation and Its Alternatives," *Proceedings of the Aristotelian Society*, vol. 19 (1919), pp. 86–124.
———, C. A. Mace, G. F. Stout, and A. C. Ewing, "Symposium: Mechanical and Teleological Causation," *Proceedings of the Aristotelian Society*, supplementary vol. 14 (1935), pp. 83–112.
Brodbeck, May, "On the Philosophy of the Social Sciences," *Philosophy of Science*, vol. 21 (1954), pp. 140–156.
Canfield, John V., "Teleological Explanation in Biology," *British Journal for the Philosophy of Science*, vol. 14 (1964), pp. 285–295.
Cohen, Jonathan, "Teleological Explanation," *Proceedings of the Aristotelian Society*, vol. 51 (1950–51), pp. 225–292.
Davis, Kingsley, "The Myth of Functional Analysis as a Special Method in Sociology and Anthropology," *American Sociological Review*, vol. 24 (1959), pp. 757–772.
Deutsch, Karl W., "Mechanism, Teleology and Mind," *Philosophy and Phenomenological Research*, vol. 12 (1951), pp. 185–223.
Ducasse, C. J., "Explanation, Mechanism and Teleology," *The Journal of Philosophy*, vol. 23 (1926), pp. 150–155. Reprinted in H. Feigl and W. Sellars (eds.), *Readings in Philosophical Analysis* (New York, 1949).
Goldstein, L. J., "The Logic of Explanation in Malinowskian Anthropology," *Philosophy of Science*, vol. 24 (1957), pp. 156–166.

Grünbaum, Adolf, "Temporally Asymmetric Principles, Parity between Explanation and Prediction, and Mechanism versus Teleology," *Philosophy of Science*, vol. 29 (1962), pp. 146–170.

Gruner, Rolf, "Teleological and Functional Explanation," *Mind*, vol. 75 (1966), pp. 516–526.

Harris, E. E., "Teleology and Teleological Explanation," *The Journal of Philosophy*, vol. 56 (1959), pp. 5–25.

Hayek, F. A., "The Facts of the Social Sciences," *Ethics*, vol. 54 (1943), pp. 1–13.

Hempel, C. G., "The Logic of Functional Analysis" in L. Gross (ed.), *Symposium on Sociological Theory* (New York, 1959), pp. 271–307.

Hofstadter, Albert, "Objective Teleology," *The Journal of Philosophy*, vol. 38 (1941), pp. 29–39.

Lehman, Hugh, "R. E. Merton's Concepts of Function and Functionalism," *Inquiry*, vol. 9 (1966), entire issue devoted to articles on functionalism.

Mace, C. A., "Symposium: Mechanical and Teleological Causation," *Proceedings of the Aristotelian Society*, supplementary vol. 14 (1935), pp. 22–45. Reprinted in H. Feigl and M. Brodbeck (eds.), *Readings in the Philosophy of Science* (New York, 1949).

Nagel, Ernest, "Teleological Explanation and Teleological Systems" in S. Ratner (ed.), *Vision and Action: Essays in Honor of Horace Kallen on His Seventieth Birthday* (New Brunswick, N.J., 1953). Reprinted in H. Feigl and M. Brodbeck (eds.), *Readings in the Philosophy of Science* (New York, 1953), pp. 537–558.

———, "A Formalization of Functionalism" in E. Nagel, *Logic Without Metaphysics* (New York, 1957), pp. 247–283.

———, "Mechanistic Explanation and Organismic Biology" in E. Nagel, *The Structure of Science* (New York, 1961), pp. 398–446.

Radcliffe-Brown, A. R., "On the Concept of Function in Social Sciences," *American Anthropologist*, vol. 37 (1935), pp. 394–402.

Rignano, Eugenio, "The Concept of Purpose in Biology," *Mind*, vol. 40 (1931), pp. 335–340.

Rosenblueth, Arturo, Norbert Wiener, and Julian Bigelow, "Behavior, Purpose, and Teleology," *Philosophy of Science*, vol. 10 (1943), pp. 18–24.

———, "The Role of Models in Science," *Philosophy of Science*, vol. 12 (1945), pp. 317–320.

Scheffler, Israel, "Thoughts on Teleology," *British Journal for the Philosophy of Science*, vol. 9 (1959), pp. 265–284. Reprinted in J. V. Canfield (ed.), *Purpose in Nature* (Englewood Cliffs, N.J., 1966).

Sorabji, Richard, "Function," *The Philosophical Quarterly*, vol. 14 (1964), pp. 289–302.

Stout, G. F., "Symposium: Mechanical and Teleological Causation," *Proceedings of the Aristotelian Society*, supplementary vol. 14 (1935), pp. 46–65.

Taylor, Richard, "Comments on and Mechanistic Conception of Purposefulness, *Philosophy of Science*, vol. 17 (1950), pp. 310–317. Reprinted in J. V. Canfield (ed.), *Purpose in Nature* (Englewood Cliffs, N.J., 1966).

D. *Explanations by Models and Analogues*

1. BOOKS AND ANTHOLOGIES

Black, Max, *Models and Metaphors* (Ithaca, N.Y., 1962).

Campbell, Norman R., *Foundations of Science* (New York, 1957).

Freudenthal, Hans (ed.), *The Concept and the Role of the Model in Mathematics and Natural and Social Sciences* (Dordrecht, Netherlands, 1962).

Hesse, Mary B., *Models and Analogies in Science* (London, 1963).

Simon, Herbert A., *Models of Man: Social and Rational* (New York, 1957).

2. ARTICLES AND CHAPTERS

Achinstein, Peter, "Models, Analogies, and Theories," *Philosophy of Science*, vol. 31 (1964), pp. 328–350.

———, "Theoretical Models," *British Journal for the Philosophy of Science*, vol. 16 (1965), pp. 102–120.

Apostel, Leo, "Towards the Formal Study of Models in the Non-Formal Sciences," *Synthese*, vol. 12 (1960), pp. 125–161.

Braithwaite, R. B., "Models in Empirical Sciences" in E. Nagel, *et al.* (eds.), *Proceedings of the Congress of the International Union for the Logic, Methodology, and Philosophy of Science* (Stanford, Calif., 1960).

Frank, Philip, "Metaphysical Interpretations of Science," *British Journal for the Philosophy of Science*, vol. 1 (1950), pp. 60–91.

Hesse, Mary B., "Models in Physics," *British Journal for the Philosophy of Science*, vol. 4 (1953), pp. 198–214.

———, "The Explanatory Function of Metaphor," *Proceedings of the 1964 International Congress for Logic, Methodology, and Philos-*

ophy of Science held in Jerusalem August 26-September 2, 1964 (Amsterdam, 1964).

Hutten, E. H., "The Role of Models in Physics," *British Journal for the Philosophy of Science*, vol. 4 (1953), pp. 284–301.

Lachman, Roy, "The Model in Theory Construction," *Phychological Review*, vol. 67 (1960), pp. 113–129. Reprinted (abridged) in Melvin H. Marx, *Theories in Contemporary Psychology* (New York, 1963).

Meadows, Paul, "Models, Systems and Science," *American Sociological Review*, vol. 22 (1957), pp. 3–9.

Meyer, Herman, "On the Heuristic Value of Scientific Models," *Philosophy of Science*, vol. 18 (1951), pp. 111–123.

Rapoport, Anatol, "Uses and Limitations of Mathematical Models in Social Sciences" in L. Gross (ed.), *Symposium on Sociological Theory* (New York, 1959).

Rosenblueth, Arturo, and Norbert Wiener, "The Role of Models in Science," *Philosophy of Science*, vol. 12 (1945), pp. 316–321.

Simon, Herbert A., "Some Strategic Considerations in the Construction of Social Science Models" in P. F. Lazarsfeld (ed.), *Mathematical Thinking in the Social Sciences* (New York, 1954).

———, and Allen Newell, "Models: Their Uses and Limitations" in L. D. White (ed.), *The State of the Social Sciences* (Chicago, 1956), pp. 66–83.

Spector, Marshall, "Models and Theories," *British Journal for the Philosophy of Science*, vol. 16 (1965), pp. 121–142.

Suppes, Patrick, "A Comparison of the Meaning and Uses of Models in Mathematics and the Empirical Sciences," *Synthese*, vol. 12 (1960), pp. 287–301.

Turner, Joseph, "Maxwell on the Method of Physical Analogy," *British Journal for the Philosophy of Science*, vol. 6 (1955), pp. 226–238.

III. Special Areas of Explanation

A. *Explanation in Physics*

1. BOOKS AND ANTHOLOGIES

Bohm, David, *Causality and Chance in Modern Physics* (London, 1957).
Born, Max, *Natural Philosophy of Cause and Chance* (London, 1949).
Bridgman, P. W., *The Logic of Modern Physics* (New York, 1927).
Campbell, Norman R., *Physics: The Elements* (Cambridge, Mass., 1920).

————, *What is Science* (London, 1920; reprinted New York, 1952).

Carnap, Rudolf, *Philosophical Foundations of Physics* (New York and London, 1966).

D'Abro, A., *The Decline of Mechanisms in Modern Physics* (New York, 1939).

Duhem, Pierre, *The Aim and Structure of Physical Theory*, tr. by P. P. Wiener (Princeton, 1954).

Hanson, Norwood R., *Patterns of Discovery* (Cambridge, 1958).

Hobson, E. W., *The Domain of Natural Science* (London, 1923).

Mach, Ernst, *The Science of Mechanics: A Critical and Historical Account of its Development*, tr. by T. J. McCormack, 2nd ed. by C. S. Peirce (Chicago, 1902).

————, *The Structure of Mechanics* (La Salle, Ill., 1942).

Margenau, Henry, *The Nature of Physical Reality* (New York, 1950).

Reichenbach, Hans, *Philosophic Foundations of Quantum Mechanics* (Berkeley, 1944).

2. ARTICLES AND CHAPTERS

Nagel, Ernest, "The Causal Character of Modern Physical Theory" in S. Baron, K. S. Pinson, and E. Nagel (eds.), *Freedom and Reason* (New York, 1951).

Rescher, Nicholas, "The Stochastic Revolution and The Nature of Scientific Explanation," *Synthese*, vol. 14 (1962), pp. 200–215.

B. *Biological Explanation*

1. BOOKS AND ANTHOLOGIES

Beckner, Morton, *The Biological Way of Thought* (Berkeley, 1968).

Blum, Harold F., *Time's Arrow and Evolution* (Princeton, 1951).

De Beer, Gavin Rylands, *Evolution* (Oxford, 1938).

Madden, Edward H. (ed.), *The Structure of Scientific Thought* (Boston, 1960).

Woodger, J. H., *Biological Principles* (New York, 1929).

2. ARTICLES AND CHAPTERS

Bonhoeffer, K. F., "Überphysikalisch-chemische Modelle von Lebensvorgängen," *Studium Generale*, vol. 1 (1948), pp. 137–143.

Canfield, John V., "Teleological Explanation in Biology," *British Journal for the Philosophy of Science*, vol. 14 (1964), pp. 285–295.

Gallie, W. B., "Explanations in History and the Genetic Sciences," *Mind*, vol. 64 (1955), pp. 161–167.

Goudge, T. A., "Some Philosophical Aspects of the Theory of Evolution," *University of Toronto Quarterly*, vol. 23 (1954), pp. 386–401.

————, "The Concept of Evolution," *Mind*, vol. 63 (1954), pp. 16–25.

Mayr, Ernst, "Cause and Effect in Biology," *Science*, vol. 134 (1961), pp. 1501–1506.

Nagel, Ernest, "Mechanistic Explanation and Organismic Biology," *Philosophy and Phenomenological Research*, vol. 11 (1950–51), pp. 327–338.

Scriven, Michael, "Explanation and Prediction in Evolutionary Theory," *Science*, vol. 130 (1959), pp. 477–482.

Sorabji, Richard, "Function," *The Philosophical Quarterly*, vol. 14 (1964), pp. 289–302.

C. *Explanation in the Social Sciences*

1. BOOKS AND ANTHOLOGIES

Argyle, Michael, *The Scientific Study. of Social Behaviour* (London, 1957).

Beard, Charles A., *The Nature of the Social Sciences* (New York, 1934).

Becker, Howard, and Alvin Boskoff (eds.), *Modern Sociological Theory* (New York, 1957).

Bidney, David, *Theoretical Anthropology* (New York, 1953).

Braybrooke, David (ed.), *Philosophical Problems of the Social Sciences* (New York, 1965).

Brown, Robert, *Explanation in Social Science* (Chicago, 1963).

Cooley, Charles H., *Sociological Theory and Social Research* (New York, 1930).

Coser, Lewis A., and Bernard Rosenberg (eds.), *Sociological Theory: A Book of Readings* (New York, 1957).

Gibson, Quentin, *The Logic of Social Enquiry* (New York and London, 1960).

Handy, Rollo, *Methodology of the Behavioral Sciences* (Springfield, Ill., 1964).

Kaufmann, Felix, *Methodology of the Social Sciences* (New York, 1964).

Lundberg. George A.. *Foundations of Sociology* (New York. 1964).

MacIver, R. M., *Social Causation* (New York, 1942).

Madge, John, *The Tools of Social Science* (New York, 1965).

Marshall, Alfred, *Principles of Economics* (New York, 8th ed. 1953).

Martindale, Don, *The Nature and Types of Sociological Theory* (Boston, 1960).

Merton, R. K., *Social Theory and Social Structure* (New York, revised ed. 1957).

Mill, John Stuart, *A System of Logic* (New York and London, 1947) originally published in 1843, Book VI, "The Logic of the Moral Sciences," abridged version in *John Stuart Mill's Philosophy of Scientific Method*, ed. by E. Nagel (New York, 1950).

Mises, Ludwig von, *Human Action: A Treatise on Economics* (New Haven, Conn., 1962).

Natanson, Maurice (ed.), *Philosophy of the Social Sciences: A Reader* (New York, 1963).

Papandreou, Andreas G., *Economics as a Science* (Chicago, 1958).

Parsons, Talcott, *The Structure of Social Action* (New York, 1937).

Robbins, Lionel, *An Essay on the Nature and Significance of Economic Science* (London, 1932).

Simon, Herbert A., *Models of Man: Social and Rational* (New York, 1957).

Timasheff, N. S., *Sociological Theory: Its Nature and Growth* (New York, 1955).

Weber, Max, *The Theory of Social and Economic Organization* (New York, 1947).

————, *The Methodology of the Social Sciences*, tr. by E. A. Shils and H. A. Finch (New York, 1949).

Winch, Peter, *The Idea of a Social Science* (London, 1958).

2. ARTICLES AND CHAPTERS

Abel, Theodore, "The Operation Called *Verstehen*," *American Journal of Sociology*, vol. 54 (1948), pp. 211–218. Reprinted in H. Feigl and M. Brodbeck (eds.), *Readings in the Philosophy of Science* (New York, 1953), pp. 677–678. Also in E. H. Madden (ed.), *The Structure of Scientific Thought* (Boston, 1960).

Albert, Ethel M., and Lewis Feuer, "Causality in the Social Sciences," *The Journal of Philosophy*, vol. 51 (1954), pp. 695–705.

Argyle, Michael, "Explanation of Social Behaviour" in M. Argyle, *The Scientific Study of Social Behaviour* (London, 1957), chap. 3.

Bartley, W. W., 3d, "Achilles the Tortoise, and Explanation in Science and History," *British Journal for the Philosophy of Science*, vol. 13 (1962–63), pp. 15–33.

Beattie, J. H. M., "Understanding and Explanation in Social Anthropology," *British Journal of Sociology*, vol. 10 (1959), pp. 45–60.

Brown, Robert, "Explanation by Laws in Social Science," *Philosophy of Science*, vol. 21 (1954), pp. 25–32.

Davis, Kingsley, "The Myth of Functional Analysis as a Special Method in Sociology and Anthropology," *American Sociological Review*, vol. 24 (1959), pp. 757–772.

Goldstein, L. J., "The Logic of Explanation in Malinowskian Anthropology," *Philosophy of Science*, vol. 24 (1957), pp. 156–166.

Grünbaum, Adolf, "Causality and the Science of Human Behavior," *American Scientist*, vol. 40 (1952), pp. 665–676. Reprinted in H. Feigl and M. Brodbeck (eds.), *Readings in the Philosophy of Science* (New York, 1953).

————, "Historical Determinism, Social Activism and Predictions in the Social Sciences," *British Journal for the Philosophy of Science*, vol. 7 (1956), pp. 236–240.

Jarvie, I. C., "Explanation in Social Science," *British Journal for the Philosophy of Science*, vol. 15 (1964), pp. 62–72. (Review discussion of *Explanation in Social Sciences* by Robert Brown.)

Lange, Oscar, "The Scope and Method of Economics," *Review of Economic Studies*, vol. 13 (1945–46), pp. 19–32. Reprinted in H. Feigl and M. Brodbeck (eds.), *Readings in the Philosophy of Science* (New York, 1953), pp. 744–754.

Nagel, Ernest, "Methodological Problems of the Social Sciences" in E. Nagel, *The Structure of Science* (New York, 1961), pp. 447–502.

————, "Explanation and Understanding in the Social Sciences" in E. Nagel, *The Structure of Science* (New York, 1961), pp. 503–546.

Olivecrona, Karl, "Is a Sociological Explanation of Law Possible?", *Theoria*, vol. 14 (1948), pp. 167–207.

Papandreou, Andreas G., "Explanation and Prediction in Economics," *Science*, vol. 129 (1958), pp. 1096–1100.

Simon, Herbert A., "Some Strategic Considerations in the Construction of Social Science Models" in P. F. Lazarsfeld (ed.), *Mathematical Thinking in the Social Sciences* (New York, 1954).

Weber, Max, "Critical Studies in the Logic of Cultural Sciences" in E. A. Shils and H. A. Finch (eds.), *The Methodology of the Social Sciences* (New York, 1949).

Zilsel, Edgar, "Physics and the Problem of Historico-Sociological Laws," *Philosophy of Science*, vol. 8 (1948), pp. 569–579. Reprinted in H. Feigl and M. Brodbeck (eds.), *Readings in the Philosophy of Science* (New York, 1953), pp. 714–722.

D. *Psychological Explanation and the Explanation of Human Behavior*

1. BOOKS AND ANTHOLOGIES

Anscombe, G. E. M., *Intention* (Ithaca, N.Y., 1957).

Berelson, Bernard, and Gary A. Steiner (eds.), *Human Behavior: An Inventory of Findings* (New York, 1964).

Bergmann, Gustav, *The Philosophy of Science* (Madison, Wisc., 1957).

Braybrooke, David, *Philosophical Problems of the Social Sciences* (New York, 1965).

Dray, William, *Laws and Explanation in History* (London, 1957).

Hull, C. L., *Principles of Behavior* (New York, 1943).

Marx, Melvin H. (ed.), *Psychological Theory: Contemporary Readings* (New York, 1951).

———, *Theories in Contemporary Psychology* (New York, 1963).

Melden, A. I., *Free Action* (London, 1961).

Parsons, Talcott, *The Structure of Social Action* (New York, 2nd ed. 1949).

———, and E. A. Shils (eds.), *Toward a General Theory of Action* (Cambridge, Mass., 1951).

Peters, R. S., *The Concept of Motivation* (New York and London, 1958).

Ryle, Gilbert, *The Concept of Mind* (New York, 1949).

Salvemini, Gaetano, *Historian and Scientist: An Essay on the Nature History and the Social Sciences* (Cambridge, Mass., 1939).

Smith, Frederick V., *The Explanation of Human Behaviour* (London, 2nd ed. 1960).

Stegmüller, Wolfgang, *Wissenschaftliche Erklärung und Begründung* (Berlin, Heidelberg, and New York, 1969).

Taylor, Charles, *The Explanation of Behaviour* (New York and London, 1964).

Taylor, Richard, *Action and Purpose* (Englewood Cliffs, N.J., 1966).

Turner, Merle B., *Philosophy and the Science of Behavior* (New York, 1967).

2. ARTICLES AND CHAPTERS

Abel, Theodore, "The Operation Called *Verstehen,*" *American Journal of Sociology,* vol. 54 (1948), pp. 211–218. Reprinted in H. Feigl and M. Brodbeck (eds.), *Readings in the Philosophy of Science* (New York, 1953) and in E. H. Madden (ed.), *The Structure of Scientific Thought* (Boston, 1960).

Alexander, Peter, "Rational Behaviour and Psychoanalytic Explanations," *Mind,* vol. 71 (1962), pp. 326–341.

Alston, William, "Wants, Actions, and Causal Explanation" (with comments by Keith Lehrer and rejoinder by Alston) in H. Castaneda (ed.), *Intentionality, Minds, and Perceptions* (Detroit, 1967).

Balmuth, Jerome, "Psychoanalytic Explanation," *Mind,* vol. 74 (1965), pp. 229–235.

Bennett, Daniel, "Action, Reason, and Purpose," *The Journal of Philosophy,* vol. 62 (1965), pp. 85–96.

Bergmann, Gustav, "On Some Methodological Problems of Psychology," *Philosophy of Science,* vol. 7 (1940), pp. 205–219. Reprinted in H. Feigl and M. Brodbeck (eds.), *Readings in the Philosophy of Science* (New York, 1953), pp. 627–636.

————, "The Logic of Psychological Concepts," *Philosophy of Science,* vol. 18 (1951), pp. 93–110.

Brandt, Richard, and Jaegwon Kim, "Wants as Explanations of Actions," *The Journal of Philosophy,* vol. 60 (1963), pp. 425–435.

Davidson, Donald, "Actions, Reasons, and Causes," *The Journal of Philosophy,* vol. 60 (1963), pp. 685–700.

Grünbaum, Adolf, "Causality and the Science of Human Behavior," *American Scientist,* vol. 40 (1952), pp. 665–676. Reprinted in H. Feigl and M. Brodbeck (eds.), *Readings in the Philosophy of Science* (New York, 1953).

————, "Science and Man," *Perspectives in Biology and Medicine,* vol. 5 (1962), pp. 483–502.

————, "Science and Man" in M. Mandelbaum, F. Gramlich, and A. Anderson (eds.), *Philosophic Problems* (New York, 1957).

————, "Free Will and Laws of Human Behavior" in W. Sellars, H. Feigl, and K. Lehrer (eds.), *Readings in Philosophical Analysis* (New York, 1953).

Hempel, Carl G., "The Concept of Rationality and the Logic of Explanation by Reasons" in C. G. Hempel, *Aspects of Explanation* (New York, 1965), pp. 463–487.

Hutten, E. H., "On Explanation in Psychology and in Physics," *British Journal for the Philosophy of Science*, vol. 7 (1956–57), pp. 73–85.

MacIntyre, A. R., "A Mistake about Causality in Social Science" in P. Laslett and W. G. Runciman (eds.), *Philosophy, Politics, and Society* (Oxford, 1962), second series B, pp. 48–70.

Madden, Edward H., "The Nature of Psychological Explanation," *Methodos*, vol. 9 (1957), pp. 53–63.

———, "Explanation in Psychoanalysis and History," *Philosophy of Science*, vol. 33 (1966), pp. 278–286.

Madell, Geoffrey, "Action and Causal Explanation," *Mind*, vol. 76 (1967), pp. 34–48.

Miller, D. L., "Meaning of Explanation," *Psychological Review*, vol. 53 (1946), pp. 241–246.

Nagel, Ernest, "Methodological Issues in Psychoanalytic Theory" in S. Hook (ed.), *Psychoanalysis, Scientific Method, and Philosophy* (New York, 1959), pp. 38–56.

Pears, D. F., "Are Reasons for Actions Causes?" in A. Stroll (ed.), *Epistemology: New Essays in the Theory of Knowledge* (New York, 1967).

Rescher, Nicholas, and Paul Oppenheim, "Logical Analysis of Gestalt Concepts," *British Journal for the Philosophy of Science*, vol. 6 (1955–56), pp. 89–106.

Skinner, B. F., "The Scheme of Behavior Explanations" in B. F. Skinner, *Science and Human Behavior* (New York, 1953). Reprinted in David Braybrooke (ed.), *Philosophical Problems of the Social Sciences* (New York, 1965), pp. 42–52.

Sutherland, N. S., "Motives and Explanations," *Mind*, vol. 68 (1959), pp. 145–159.

Taylor, Richard, "Comments on and Mechanistic Conception of Purposefulness," *Philosophy of Science*, vol. 17 (1950), pp. 310–317. Reprinted in J. V. Canfield (ed.), *Purpose in Nature* (Englewood Cliffs, N.J., 1966).

E.   Historical Explanation

1.   BOOKS AND ANTHOLOGIES

NOTE: For a comprehensive bibliography of the field see Patrick Gardiner, *Theories of History* (New York, 1959).

Berlin, Isaiah, *Historical Inevitability* (New York and London, 1954).

Danto, Arthur C., *Analytical Philosophy of History* (Cambridge, 1965).
Dray, William H., *Laws and Explanation in History* (London, 1957).
———, (ed.), *Philosophical Analysis and History* (New York, 1966).
Engels, Friedrich, *On Historical Materialism* (New York, n.d.).
Gallie, W. B., *Philosophy and the Historical Understanding* (London, 1964).
Gardiner, Patrick, *The Nature of Historical Explanation* (London, 1952).
———, *Theories of History* (New York, 1959).
White, Morton, *Foundations of Historical Knowledge* (New York, 1965).

2. ARTICLES AND CHAPTERS

Bartley, W. W., 3d., "Achilles, the Tortoise, and Explanation in Science and History," *British Journal for the Philosophy of Science*, vol. 13 (1962), pp. 15–33.
Cornforth, K., "Symposium: Explanation in History," *Proceedings of the Aristotelian Society*, supplementary vol. 14 (1935), pp. 123–141.
Danto, Arthur C., "On Explanations in History," *Philosophy of Science*, vol. 23 (1956), pp. 15–30.
Donagan, Alan, "Explanation in History," *Mind*, vol. 66 (1957), pp. 145–164. Reprinted in P. Gardiner (ed.), *Theories of History* (New York, 1959).
———, "Historical Explanations: The Popper-Hempel Theory Reconsidered," *History and Theory*, vol. 4 (1964), pp. 3–26. Reprinted in W. H. Dray (ed.), *Philosophical Analysis and History* (New York, 1966).
Dray, William H., "Explanatory Narrative in History," *The Philosophical Quarterly*, vol. 4 (1954), pp. 15–27.
———, " 'Explaining What' in History" in P. Gardiner (ed.), *Theories of History* (New York, 1959), pp. 403–408.
———, "The Historical Explanation of Actions Reconsidered" in S. Hook (ed.), *Philosophy and History* (New York, 1963), pp. 105–135.
Frankel, Charles, "Explanation and Interpretation in History" in P. Gardiner (ed.), *Theories of History* (New York, 1959), pp. 408–427.
Gallie, W. B., "Explanation in History and the Genetic Sciences," *Mind*, vol. 64 (1955), pp. 160–180. Reprinted in P. Gardiner (ed.), *Theories of History* (New York, 1959).
Ginsberg, Morris, "The Character of an Historical Explanation," *Proceedings of the Aristotelian Society*, supplementary vol. 21 (1947), pp. 69–77.

Goudge, T. A., "Causal Explanation in Natural History," *British Journal for the Philosophy of Science*, vol. 9 (1958), pp. 194–202.

Grünbaum, Adolf, "Historical Determinism, Social Activism, and Prediction in the Social Sciences," *British Journal for the Philosophy of Science*, vol. 7 (1956), pp. 236–240.

Hempel, Carl G., "The Function of General Laws in History," *The Journal of Philosophy*, vol. 39 (1942), pp. 35–48. Reprinted in H. Feigl and W. Sellars (eds.), *Readings in Philosophical Analysis* (New York, 1949).

———, "Explanation in Science and in History" in R. G. Colodny (ed.), *Frontiers of Science and Philosophy* (Pittsburgh, 1962), pp. 7–34.

MacIver, A. M., "The Character of an Historical Explanation," *Proceedings of the Aristotelian Society*, supplementary vol. 21 (1947), pp. 33–50.

Madden, Edward H., "Explanation in Psychoanalysis and History," *Philosophy of Science*, vol. 33 (1966), pp. 278–286.

Mandelbaum, Maurice, "Historical Explanation: The Problem of 'Covering Laws,'" *History and Theory*, vol. 1 (1961), pp. 229–242.

Melden, A. I., "Historical Objectivity, A 'Noble Dream'?" *Journal of General Education*, vol. 7 (1952), pp. 17–24.

Nagel, Ernest, "Problems in the Logic of Historical Inquiry" in E. Nagel, *The Structure of Science* (New York, 1961), pp. 547–606.

Newman, Fred, "Explanation Sketches," *Philosophy of Science*, vol. 32 (1965), pp. 168–172.

Nilson, S. S., "Mechanics and Historical Laws," *The Journal of Philosophy*, vol. 48 (1951), pp. 201–211.

Passmore, John, "Law and Explanation in History," *The Australian Journal of Politics and History*, vol. 4 (1958), pp. 269–276.

———, "Explanation in Everyday Life, in Science, and in History," *History and Theory*, vol. 2 (1962–63), pp. 105–123.

Pitt, Jack, "Generalizations in Historical Explanation," *The Journal of Philosophy*, vol. 56 (1959), pp. 578–586.

Rescher, Nicholas, and Carey B. Joynt, "The Problem of Uniqueness in History," *History and Theory*, vol. 1 (1961), pp. 150–162. Reprinted in George H. Nadel (ed.), *Studies in the Philosophy of History* (New York, 1965).

Scriven, Michael, "Truisms as the Grounds for Historical Explanation" in P. Gardiner (ed.), *Theories of History* (New York, 1959), pp. 443–475.

———, "Causes, Connections and Conditions in History" in W. H. Dray (ed.), *Philosophical Analysis and History* (New York, 1966).

Strong, E. W., "Fact and Understanding in History," *The Journal of Philosophy*, vol. 44 (1947), pp. 617–625.

————, "Criteria of Explanation in History," *The Journal of Philosophy*, vol. 49 (1952), pp. 57–67.

Walsh, W. H., "The Character of an Historical Explanation," *Proceedings of the Aristotelian Society*, supplementary vol. 21 (1947), pp. 51–68.

Watkins, J. W. N., "Ideal Types and Historical Explanation," *British Journal for the Philosophy of Science*, vol. 3 (1952), pp. 22–43. Reprinted in H. Feigl and M. Brodbeck (eds.), *Readings in the Philosophy of Science* (New York, 1953), pp. 723–743.

————, "Historical Explanation in the Social Sciences," *British Journal for the Philosophy of Science*, vol. 8 (1957), pp. 104–116.

Weingartner, R. H., "The Quarrel about Historical Explanation," *The Journal of Philosophy*, vol. 58 (1961), pp. 29-45.

White, Morton, "Toward an Analytic Philosophy of History" in M. Farber (ed.), *Philosophic Thought in France and the United States* (Buffalo, 1950).

————, "Historical Explanation," *Mind*, vol. 52 (1943), pp. 212–229.

Zilsel, Edgar, "Physics and the Problem of Historico-Sociological Laws," *Philosophy of Science*, vol. 8 (1941), pp. 567–579. Reprinted in H. Feigl and M. Brodbeck (eds.), *Readings in the Philosophy of Science* (New York, 1953), pp. 714–722.

IV. Special Topics

A. *Explanation versus Prediction*

1. BOOKS AND ANTHOLOGIES

Goodman, Nelson, *Fact, Fiction, and Forecast* (Cambridge, Mass., 1955).

Grünbaum, Adolf, *Philosophy of Space and Time* (New York, 1964).

Reichenbach, Hans, *Nomological Statements and Admissible Operations* (Amsterdam, 1954).

Stegmüller, Wolfgang, *Wissenschaftliche Erklärung und Begründung* (Berlin, Heidelberg, and New York, 1969).

2. ARTICLES AND CHAPTERS

Brodbeck, May, "Explanations, Predictions, and 'Imperfect' Knowledge" in H. Feigl and G. Maxwell (eds.), *Minnesota Studies in the*

Philosophy of Science, vol. 3 (Minneapolis, 1962), pp. 231–272.

Buck, Roger, "Reflexive Predictions," *Philosophy of Science*, vol. 30 (1963), pp. 359–369.

Carnap, Rudolf, "The Methodological Character of Theoretical Concepts" in H. Feigl, M. Scriven, and G. Maxwell (eds.), *Minnesota Studies in the Philosophy of Science*, vol. 2 (Minneapolis, 1958).

Coffa, J. A., "Deductive Predictions," *Philosophy of Science*, vol. 35 (1968), pp. 279–283.

Grünbaum, Adolf, "Temporally Asymmetric Principles, Parity between Explanation and Predication and Mechanism versus Teleology," *Philosophy of Science*, vol. 29 (1962), pp. 146–170. Reprinted in H. Kyburg (ed.), *Induction: Some Current Issues* (Middletown, Conn., 1963), in B. Baumrin (ed.), *Philosophy of Science: The Delaware Seminar*, vol. 1 (New York, 1963), pp. 57–96.

———, "Reflexive Predictions: Comments on Professor Roger Buck's Paper," *Philosophy of Science*, vol. 30 (1963), pp. 370–372.

———, "The Scientific Assertibility of Statements about the Past and about the Future" to appear in vol. 4 of the University of Pittsburgh series in the Philosophy of Science, ed. by R. G. Colodny.

Grünberg, Emile, and Franco Modigliani, "Reflexive Prediction," *Philosophy of Science*, vol. 32 (1965), pp. 173–174.

Hanson, Norwood R., "On the Symmetry between Explanation and Prediction," *The Philosophical Review*, vol. 68 (1959), pp. 349–358.

Rescher, Nicholas, "On Prediction and Explanation," *British Journal for the Philosophy of Science*, vol. 8 (1958), pp. 281–290.

———, "Discrete State Systems, Markov Chains and Problems in the Theory of Scientific Explanation and Prediction," *Philosophy of Science*, vol. 39 (1963), pp. 325–345.

Scheffler, Israel, "Explanation, Prediction, and Abstraction," *British Journal for the Philosophy of Science*, vol. 7 (1957), pp. 293–309. Reprinted in A. Danto and S. Morgenbesser (eds.), *Philosophy of Science* (New York, 1960).

Scriven, Michael, "Definitions, Explanations, and Theories" in H. Feigl, G. Maxwell, and M. Scriven (eds.), *Minnesota Studies in the Philosophy of Science*, vol. 2 (Minneapolis, 1958), pp. 99–195.

———, "Explanation and Prediction in Evolutionary Theory," *Science*, vol. 130 (1959), pp. 477–482.

———, "Explanations, Predictions, and Laws" in H. Feigl and G. Maxwell (eds.), *Minnesota Studies in the Philosophy of Science*, vol. 3 (Minneapolis, 1962), pp. 170–230.

———, "The Temporal Asymmetry of Explanations and Predictions" in

B. Baumrin (ed.), *Philosophy of Science: The Delaware Seminar*, vol. 1 (New York, 1963), pp. 97–105.

B. *The Role of Laws and Theories in Explanation*

1. BOOKS AND ANTHOLOGIES

Barker, S. F., *Induction and Hypothesis* (New York, 1957).
Goodman, Nelson, *Fact, Fiction, and Forecast* (Cambridge, Mass., 1955).
Nagel, Ernest, *The Structure of Science* (New York, 1961).
Peirls, R. E., *The Laws of Nature* (New York, 1956).
Popper, Karl R., *The Logic of Scientific Discovery* (London, 1959).
Reichenbach, Hans, *Experience and Prediction* (Chicago, 1938).
————, *Nomological Statements and Admissible Operations* (Amsterdam, 1954). See the review by C. G. Hempel in *The Journal of Symbolic Logic*, vol. 26 (1956), pp. 50–54.
Stegmüller, Wolfgang, *Wissenschaftliche Erklärung und Begründung* (Berlin, Heidelberg, and New York, 1969).

2. ARTICLES AND CHAPTERS

Alexander, H. Gavin, "General Statements as Rules of Inference" in H. Feigl, G. Maxwell and M. Scriven (eds.), *Minnesota Studies in the Philosophy of Science*, vol. 2 (Minneapolis, 1958).
Barker, S. F., "The Role of Simplicity in Explanation" in H. Feigl, M. Scriven, and G. Maxwell (eds.), *Minnesota Studies in the Philosophy of Science*, vol. 2 (Minneapolis, 1958).
Boltzmann, Ludwig, "Theories as Representations" in *Die Grundprinzipien und Grundgleichungen der Mechanik, I* (Leipzig, 1905). Reprinted in A. Danto and S. Morgenbesser (eds.), *Philosophy of Science*, tr. by R. Weingartner (New York, 1960).
Brodbeck, May, "Explanation, Prediction and 'Imperfect' Knowledge" in H. Feigl and G. Maxwell (eds.), *Minnesota Studies in the Philosophy of Science*, vol. 3 (Minneapolis, 1962), pp. 231–272.
Campbell, Norwood R., "The Structure of Theories" in N. R. Campbell, *Physics: The Elements* (Cambridge, Mass., 1920). Reprinted in H. Feigl and M. Brodbeck (eds.), *Readings in the Philosophy of Science* (New York, 1953).
Chisholm, Roderick M., "Law Statements and Counterfactual Inference," *Analysis*, vol. 15 (1955), pp. 97–105.

Grünbaum, Adolf, "Law and Convention in Physical Theory" in H. Feigl and G. Maxwell (eds.), *Current Issues in the Philosophy of Science* (New York, 1961).

Hempel, Carl G., "The Function of General Laws in History," *The Journal of Philosophy*, vol. 39 (1942), pp. 35–48.

———, "Explanation and Prediction by Covering Laws" in B. Baumrin (ed.), *Philosophy of Science: The Delaware Seminar*, vol. 1 (New York, 1963).

———. "Inductive Inconsistencies." *Synthese*. vol. 12 (1960). pp. 439–469. Reprinted in C. G. Hempel. *Aspects of Scientific Explanation* (New York and London. 1965).

Hesse, Mary B., "Theories, Dictionaries, and Observation," *British Journal for the Philosophy of Science*, vol. 9 (1958), pp. 12–28.

Jeffrey, Richard, "Valuation and Acceptance of Scientific Hypotheses," *Philosophy of Science*, vol. 23 (1956), pp. 237–246.

Körner, Stephen, "On Laws of Nature," *Mind*, vol. 62 (1963), pp. 218–229.

Levi. Isaac. "Must the Scientist Make Value Judgements?" *The Journal of Philosophy*. vol. 57 (1960). pp. 345–357.

———, "Deductive Cogency in Inductive Inference," *The Journal of Philosophy*, vol. 62 (1965), pp. 68–77.

Lundberg, G. A., "The Concept of Law in the Social Sciences," *Philosophy of Science*, vol. 5 (1938), pp. 189–203.

Mandelbaum, Maurice, "Societal Laws," *British Journal for the Philosophy of Science*, vol. 8 (1957), pp. 211–224.

Nagel, Ernest, "The Logical Character of Scientific Laws" in E. Nagel, *The Structure of Science* (New York, 1961), pp. 47–78.

———, "Experimental Laws and Theories" in E. Nagel, *The Structure of Science* (New York, 1961), pp. 79–105.

———, "The Cognitive Status of Theories" in E. Nagel, *The Structure of Science* (New York, 1961), pp. 106–152.

———, "The Reduction of Theories" in E. Nagel, *The Structure of Science* (New York, 1961), pp. 336–397.

Passmore, John, "Prediction and Scientific Law," *Australasian Journal of Philosophy*, vol. 24 (1946), pp. 1–33.

Rudner, Richard, "The Scientist Qua Scientist Makes Value Judgements," *Philosophy of Science*, vol. 20 (1953), pp. 1–6.

Schlick, Moritz, "Are Natural Laws Conventions?" in H. Feigl and M. Brodbeck (eds.), *Readings in the Philosophy of Science* (New York, 1953).

Scriven, Michael, "Definitions, Explanations, and Theories" in H. Feigl,

M. Scriven, and G. Maxwell (eds.), *Minnesota Studies in the Philosophy of Science*, vol. 2 (Minneapolis, 1958), pp. 99–195.

———, "Explanations, Predictions, and Laws" in H. Feigl and G. Maxwell (eds.), *Minnesota Studies in the Philosophy of Science*, vol. 3 (Minneapolis, 1962), pp. 170–230.

Stegmüller, Wolfgang, "Der Bergriff des Naturgesetzes." *Studium Generale*, vol. 19 (1966), pp. 649–657.

Wigner, Eugene, "Events, Laws of Nature, and Invariance Principles," *Science*, vol. 145 (1964), pp. 995–998.

C. Counterfactuals and Scientific Explanation

1. BOOKS AND ANTHOLOGIES

Braithwaite, R. B., *Scientific Explanation* (Cambridge, 1953).

Goodman, Nelson, *Fact, Fiction, and Forecast* (Cambridge, Mass., 1955).

Johnson, W. E., *Logic* (Cambridge, Mass., 1924).

Kneale, William, *Probability and Induction* (Oxford, 1949).

Lewis, C. I., *An Analysis of Knowledge and Valuation* (La Salle, Ill., 1946).

Nagel, Ernest, *The Structure of Science* (London, 1961).

Pap, Arthur, *An Introduction to the Philosophy of Science* (New York, 1962).

Popper, Karl R., *The Logic of Scientific Discovery* (London, 1959).

Reichenbach, Hans, *Nomological Statements and Admissible Operations* (Amsterdam, 1954).

Rescher, Nicholas, *Hypothetical Reasoning* (Amsterdam, 1964).

Ryle, Gilbert, *The Concept of Mind* (New York, 1949).

Stegmüller, Wolfgang, *Wissenschaftliche Erklärung und Begründung* (Berlin, Heidelberg, and New York, 1969).

Wright, G. H. von, *Logical Studies* (London, 1957).

2. ARTICLES AND CHAPTERS

Burks, A. W., "Dispositional Statements," *Philosophy of Science*, vol. 22 (1955), pp. 175–193.

Carnap, Rudolf, "Testability and Meaning," *The Journal of Philosophy*, vol. 3 (1936), pp. 419–471; vol. 4 (1937), pp. 1–40.

Chisholm, Roderick M., "The Contrary-to-fact Conditional," *Mind*, vol. 55 (1946), pp. 289–307. Amended in H. Feigl and W. Sellars (eds.),

Readings in Philosophical Analysis (New York, 1949).

———, "Law Statements and Counterfactual Inference," *Analysis*, vol. 15 (1955), pp. 97–105.

Cooley, John C., "Professor Goodman's 'Fact, Fiction, and Forecast,'" *The Journal of Philosophy*, vol. 54 (1957), pp. 293–311.

Diggs, B. J., "Counterfactual Conditionals," *Mind*, vol. 61 (1952), pp. 513–527.

Goodman, Nelson, "The Problem of Counterfactual Conditionals," *The Journal of Philosophy*, vol. 44 (1947), pp. 113–128. Reprinted in L. Linsky (ed.), *Semantics and the Philosophy of Language* (Urbana, 1952), with minor changes in N. Goodman, *Fact, Fiction, and Forecast* (Cambridge, Mass., 1955).

Hampshire, Stuart, "Subjunctive Conditionals," *Analysis*, vol. 9 (1948), pp. 9–14. Reprinted in M. MacDonald (ed.), *Philosophy and Analysis* (Oxford, 1955).

Hempel, Carl G., "Studies in the Logic of Confirmation," *Mind*, vol. 54 (1945), pp. 1–26, 97–121.

Hiz, Henry, "On the Inferential Sense of Contrary-to-Fact Conditionals," *The Journal of Philosophy*, vol. 48 (1951), pp. 586–587.

Kneale, William, "Natural Laws and Contrary-to-Fact Conditionals," *Analysis*, vol. 10 (1950), pp. 123–125. Reprinted in M. MacDonald (ed.), *Philosophy and Analysis* (Oxford, 1955).

Mackie, J. L., "Counterfactuals and Causal Laws" in R. J. Butler (ed.), *Analytical Philosophy* (New York, 1962), pp. 66–80.

Pap, Arthur, "Disposition Concepts and Extensional Logic" in H. Feigl, M. Scriven, and G. Maxwell (eds.), *Minnesota Studies in the Philosophy of Science*, vol. 2 (Minneapolis, 1958).

Popper, Karl R., "A Note on Natural Laws and So-Called Contrary-to-Fact Conditionals," *Mind*, vol. 58 (1949), pp. 62–66.

Rescher, Nicholas, "Belief-Contravening Suppositions," *The Philosophical Review*, vol. 70 (1961), pp. 176–196.

Schneider, Erna F., "Recent Discussions of Subjunctive Conditionals," *Review of Metaphysics*, vol. 6 (1952), pp. 623–647.

Sellars, Wilfrid, "Counterfactuals, Dispositions, and the Causal Modalities" in H. Feigl, M. Scriven, and G. Maxwell (eds.), *Minnesota Studies in the Philosophy of Science*, vol. 2 (Minneapolis, 1958), pp. 225–308.

Stegmüller, Wolfgang, "Conditio Irrealis, Dispositionen, Naturgesetze und Induktion," *Kantstudien*, vol. 50 (1958–59), pp. 363–390.

Storer, Thomas, "On Defining 'Soluble,'" *Analysis*, vol. 11 (1950–51), pp. 134–137.

Walters, R. S., "The Problem of Counterfactuals," *Australasian Journal of Philosophy*, vol. 39 (1961), pp. 30–46.

Weinberg, Julius B., "Contrary-to-Fact Conditionals," *The Journal of Philosophy*, vol. 48 (1951), pp. 17–22.

Will, F. L., "The Contrary-to-Fact Conditional," *Mind*, vol. 56 (1947), pp. 236–249.

Index of Names

Note: This Index is exclusive of the Bibliography.